SPIDERS
AND THEIR KIN

a Golden Guide® from St. Martin's Press
by
HERBERT W. LEVI
Museum of Comparative Zoology
Harvard University
and
LORNA R. LEVI

Under the editorship of
HERBERT S. ZIM

Illustrated by
NICHOLAS STREKALOVSKY

The revision prepared by
the authors
and Jonathan P. Latimer
and
Karen Stray Nolting

St. Martin's Press 🦋 New York

FOREWORD

This small guide to spiders and their near relatives introduces the various groups and shows their great diversity. Accurate species identification is often a problem even for specialists, and while the groups treated in this guide are widespread, some of the species illustrated have a limited distribution. If they are not found where you live, perhaps you will find spiders that are similar. The scope of the book is broad enough to make it useful in Europe and on other continents.

The book would have been impossible without the help of numerous friends and colleagues. Among those who read early drafts of the text were Harriet Exline Frizzell, W. J. Gertsch, O. Kraus, Nell B. Causey, and R. Crabill. Mr. N. Strekalovsky made the illustrations, often handicapped by limitations of live source material. Superb color slides of European spiders were made available by J. Pötzsch; slides of many uncommon species were loaned by H. K. Wallace. We sincerely thank all these and also the many who provided living animals, color photographs, determinations of unfamiliar animals, or help with the text: J. W. Abalos, G. Anastos, C. R. Baird, E. W. Baker, J. Beatty, A. R. Brady, P. Bonnet, Stephanie Cannon, Nell B. Causey, J. Cokendolpher, J. A. Coddington, B. Condé, J. A. L. Cooke, F. A. Coyle, J. Davis, C. D. Dondale, L. C. Drew, W. Eberhard, G. B. Edwards, T. Eisner, G. S. Fichter, B. T. Gardner, G. P. Ginsburg, L. Glatz, B. Heydemann, R. L. Hoffman, H. Homann, J. Jass, J. Kaspar, B. J. Kaston, B. Klausmeier, H. Klingel, G. M. Kohls, R. König, D. H. Lamore, Z. Maretić, J. Martens, M. Melchers, Rodger Mitchell, W. B. Muchmore, M. H. Muma, F. Papi, B. Patterson, N. I. Platnick, J. Rafalski, J. Reddell, Jonathan Reiskind, V. D. Roth, J. H. P. Sankey, P. San Martin, P. Stough, W. D. Sill, H. Stahnke, T. W. Suman, D. W. Sissom, W. A. Shear, Paolo Tongiorgi, M. W. Tyler, J. D. Unzicker, M. Vachon, A. A. Weaver, G. C. Wheeler, P. Witt, T. A. Woolley.

H.L.
L.L.

ISBN 1-58238-156-9

CONTENTS

About 35,000 species of spiders have been named so far, representing what is believed to be about one-fifth the total. Some 3,000 kinds are known from Europe; 3,500 from less-studied N.A. The 700 species of spiders found in New York and New England about equals the species of birds breeding in N.A. north of Mexico.

Spiders are members of the phylum Arthropoda, the large group of animals with jointed legs and a hard outer skeleton. They belong, more specifically, to the class Arachnida, which includes animals with four pairs of legs, no antennae or wings, and only two body regions—a cephalothorax and an abdomen. Arachnids and two smaller marine arthropod groups (pp. 6–7) form the subphylum Chelicerata. These arthropods all possess chelicerate jaws (pp. 8, 20, 26) which sometimes are modified into pincers as in windscorpions (p. 118–119) or into piercing stylets, as in some mites (p. 134).

All other arthropods have antennae and mandibles that work against each other. They are placed in six classes. These include the insects (class Insecta), which have three body regions, three pairs of legs, one pair of antennae, and often wings; the crustaceans (class Crustacea), mainly water-dwellers—the crabs, lobsters, shrimp, barnacles, and water fleas; and the myriapods: centipedes (class Chilopoda) and millipedes (class Diplopoda); and classes Symphyla and Pauropoda found in habitats like those of spiders.

In spiders, the abdomen is attached to the cephalothorax by a narrow stalk; in scorpions, harvestmen, and mites, the attachment is broad. Spiders usually have eight simple eyes, variously arranged, and some have acute vision. Scorpions have both median eyes and (usually) lateral eyes; harvestmen, only median eyes; pseudoscorpions, lateral eyes or none. Long setae (hairs), sensitive to vibration, air movements and sound, occur on the legs of some spiders and on the pedipalps of scorpions and pseudoscorpions.

In spiders the abdomen shows little or no segmentation, but segments are distinct in scorpions. The segmented spiders, suborder Mesothelae, family Liphistiidae (p. 7), are an exception. These spiders live in burrows in the soil and are only found in East Asia. In the other suborders of spiders, Mygalomorphae (p. 20) and Araneomorphae (p. 26), vestiges of the segmentation thought to characterize ancestral forms, is reflected externally by the pattern on the back of the abdomen and sometimes by the presence of several hard plates (sclerites); internally by the muscle arrangement and structure of the heart.

This book treats the land arthropods other than insects. Of these, the spiders and mites, both of the class Aarchnida, are the most abundant. Mites are mostly microscopic and difficult to study, hence are given little attention here. The graphs below show the number of species in major groups of arthropods and arachnids.

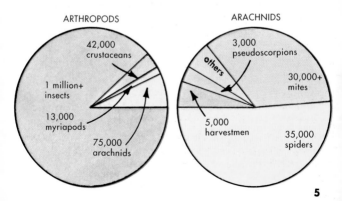

ARTHROPODS

42,000 crustaceans

1 million+ insects

13,000 myriapods

75,000 arachnids

ARACHNIDS

3,000 pseudoscorpions

others

30,000+ mites

5,000 harvestmen

35,000 spiders

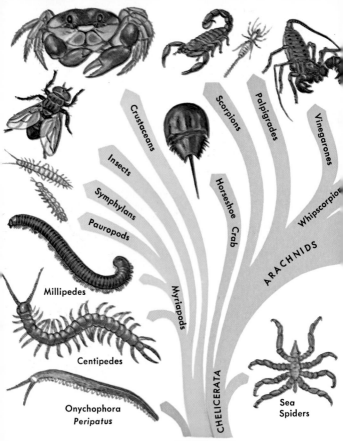

Crustaceans

Scorpions

Palpigrades

Vinegarones

Insects

Horseshoe Crab

Symphylans

Pauropods

Whipscorpions

Millipedes

ARACHNIDS

Myriapods

Centipedes

Onychophora
Peripatus

CHELICERATA

Sea
Spiders

ARTHROPODS

ARTHROPODS evolved from marine segmented worms. Fossils do
not reveal whether they evolved from different stocks or all from
one group. Onychophorans, represented by the many-legged, soft-
bodied *Peripatus* mainly of the Southern Hemisphere, are perhaps
similar to some ancestor. The groups of arthropods and their prob-
able relationships are shown above.

The most wormlike arthropods are certain centipedes (p. 142)

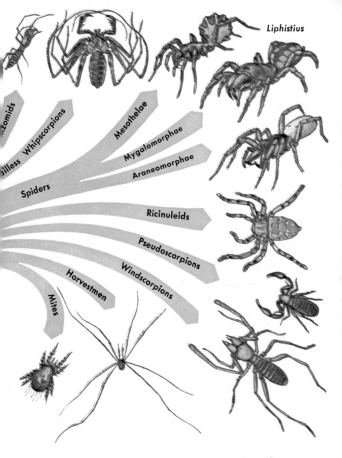

Liphistius

...zomids

...illess Whipscorpions

Mesothelae

Mygalomorphae

Spiders

Araneomorphae

Ricinuleids

Pseudoscorpions

Windscorpions

Harvestmen

Mites

that consist of series of similar segments. In other arthropods, groups of segments have become specialized. In insects, one group of segments forms the head; another, the thorax; a third, the abdomen.

Two marine groups related to the arachnids are included in the subphylum Chelicerata: horseshoe crabs, which live only on the east coasts of Asia and N.A.; and sea spiders, which are slow marine creatures that feed on hydroids, anemones and other sea animals.

THE SPIDER'S BODY consists of a cephalothorax, covered by a carapace (shield), and an abdomen. Four pairs of legs are attached to the cephalothorax. The legs end in either two or three claws, varying with the family. Nearly all spiders have eight simple eyes. Their arrangement, important in identifying families, is shown in black-and-white diagrams with family descriptions. The shape of the carapace is commonly distinctive, too.

The cephalothorax (combined head and thorax) contains the brain, venom glands (p. 16), and stomach. In the abdomen (p. 13) are the heart, digestive tract, reproductive organs, lungs and respiratory tracheae, and silk glands. The two parts are connected by a thin stalk, the pedicel, through which pass the aorta, intestine, nerve cord, and some muscles. Spinnerets (usually six) issue strands of silk through tiny spigots. Between the front pair in some spiders is the colulus, its function unknown. In cribellate spiders (p. 106), the cribellum is here.

THE JAWS, or chelicerae, open forward in Mygalomorphae (p. 20) and in the Mesothelae (p. 7), and to the sides in other spiders. Spider jaws are tipped by fangs, with a duct from a venom gland opening at the end of each. In front of the labium (lower lip) is the mouth, its opening covered by the labrum (upper lip).

Spiders feed on living prey, which may be paralyzed or killed with venom. Juices from the digestive glands liquefy the prey before it is sucked into the mouth by the stomach's pumping action. Spiders with few teeth on their jaws may suck out the insides of prey and discard the empty shell.

PEDIPALPS, between the jaws and the first legs, are small and leglike in females and in young spiders. In males, the tip is enlarged. Before searching for a female, the male deposits a drop of sperm on a special web, then draws it into the palp.

In mating, the sperm is transferred by inserting the palp into an opening on the underside of the female's abdomen. In most spiders, this opening is on a hard plate, the epigynum, just in front of a slit (gonopore) through which the eggs pass. Some, called haplogyne spiders (pp. 26–30), lack an epigynum; the palpus is inserted directly into the gonopore.

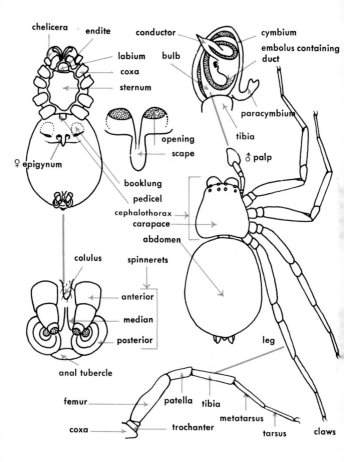

- chelicera
- endite
- labium
- coxa
- sternum
- conductor
- bulb
- cymbium
- embolus containing duct
- paracymbium
- tibia
- ♂ palp
- ♀ epigynum
- opening
- scape
- booklung
- pedicel
- cephalothorax
- carapace
- abdomen
- colulus
- spinnerets
- anterior
- median
- posterior
- leg
- anal tubercle
- femur
- patella
- tibia
- metatarsus
- coxa
- trochanter
- tarsus
- claws

EXTERNAL FEATURES shown above are used in the text to describe families and species.

Measurements given in this book are of the adult spider's approximate body length, excluding legs and jaws. They are not measurements of leg span. All spiders show individual and regional differences in size. The sign ♀ is used for female, ♂ for male.

COURTSHIP by the adult male begins after his palp is filled with sperm (p. 8) and he has found a female.

Some hunting spiders locate mates by finding and following the draglines (p. 15) laid down by mature females of the same species. Experiments have demonstrated that male web spiders can often tell by touching the web whether it contains a mature female. Male orb weavers and other web spiders with poor vision announce their approach by plucking the strands of the female's web. Others stroke and tap the female cautiously. Spiders with good vision, such as wolf spiders (below) and the brightly colored jumping spiders, dance and wave their legs before their mates. A nursery web spider (p. 78) presents his mate with a fly before mating.

A female does not ordinarily feast on her mate, as many people believe, but males usually die soon after mating. Some male and female sheet web spiders (p. 46) live together in the same web.

After a week or more, the mated female deposits her eggs in a silken sac (p. 14). Some species make several egg sacs, each containing several hundred eggs. Species that take care of their eggs or young usually produce fewer eggs. Weeks later, or sometimes not until the following spring, the young spiderlings emerge.

Courtship of wolf spiders
Pardosa nigriceps

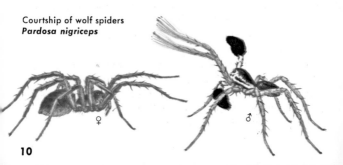

♀

♂

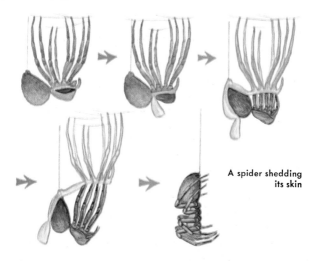

A spider shedding
its skin

GROWTH of a spider requires shedding its exoskeleton, usually 4 to 12 times before maturity. Female mygalomorph spiders (p. 20) continue to molt once or twice a year through their long adult lives.

Before a spider sheds, the inside layers of its skeleton are digested. The remaining skeleton then tears more easily. As molting begins, increased blood pressure causes the skeleton to tear at the front edge, continuing around the carapace, which then lifts off; the skin of the abdomen splits. A pumping motion lowers and raises leg spines, making the old skin slip over the flexible new legs. In the process of molting, a previously lost leg may be replaced by a new, smaller leg.

Most spiders live one or two seasons. Mygalomorphs do not mature for several years; the males live less than a year thereafter, but the females may live up to 20 years. Some primitive araneomorphs (such as *Sicarius, Loxosceles,* and *Kukulcania*) may live 5 to 10 years.

ENEMIES of spiders include other spiders (pp. 50, 51) and some kinds of insects and birds. These enemies help to control spider populations, which are affected also by parasites and availability of food. The spiders in turn control to some degree the abundance of the prey they feed on, usually insects.

WASPS of some species prey only on spiders belonging to particular families or genera.

Below, a female wasp, *Anoples fuscus*, is stinging a wolf spider, *Trochosa terricola* (1). (Rarely the spider is the victor in these bat-tles.) The wasp then carries the spider to her previously excavated burrow (2) and lays an egg on it (3). The wasp larva feeds on the paralyzed spider (4), eventually to pupate and metamorphose into an adult.

Pötzsch

SILK produced by spiders is used in many ways (pp. 14-15). Pseudoscorpions (p. 120), spider mites (p. 135), most centipedes (p. 142), and some millipedes (p. 148) also produce silk but only for mating or for egg and larval chambers. The caterpillars of many moths spin silk for their cocoons.

CHEMICALLY silk is a fibrous protein (fibroin), insoluble in water. It comes from spigots of the spinnerets in liquid form and hardens immediately, polymerizing as it is pulled out. Silk may stretch as much as one-fourth its length before breaking, and the silk of *Nephila* (p. 65) is the strongest natural fiber known. Spider silk is not used commercially, as the predatory habit of spiders makes it difficult to rear them in large numbers. Web spiders produce different types of silk from four to seven abdominal glands: viscid or sticky silk from some, web frame threads from others, egg sac silk from still others.

CLAWS on the tips of the legs are used by the spider to handle silk. The claws pivot back and forth, holding the silk between middle claw and two flexible setae (accessory claws). What prevents silk from sticking is not known. Spiders that build webs and walk on the silk threads have three claws on each leg. However, many hunting spiders have only two claws, for in place of the middle claw is a tuft of flattened hairs. Some also have a brush of hairs (scopula) under each leg's last segment. The claw tuft adheres to the water film covering most surfaces, permitting a spider to walk on smooth areas.

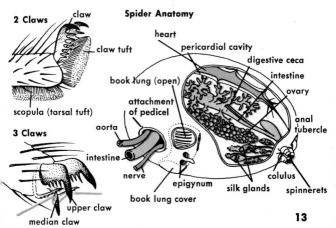

Spider Anatomy

2 Claws — claw, claw tuft, scopula (tarsal tuft)

3 Claws — upper claw, median claw, intestine, nerve

heart, pericardial cavity, digestive ceca, intestine, ovary, anal tubercle, book lung (open), attachment of pedicel, aorta, epigynum, book lung cover, silk glands, colulus, spinnerets

13

Zora spinimana, European Zoridae, 6 mm (0.2"), guarding egg sac

MANY USES OF SILK have evolved. Most spiders make silken egg cases, often spherical but sometimes flattened discs or stalked. Some species use silk to make a nursery for spiderlings (p. 78). Many hide in silk tunnels or use silk to line their burrows, or for trapdoors (p. 22). Prey may be caught in webs, or snares, then wrapped (p. 53).

EGG SAC of *Argiope bruennichi* is shown below. The eggs are first stuck to a silk platform, then covered with threads. After they are wrapped in loose silk, a final cover of dense, colored silk is added. *Argiope* suspends the egg sac from vegetation.

Pötzsch

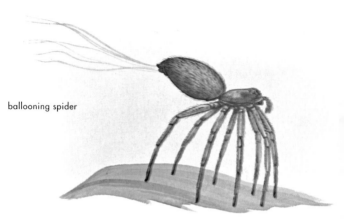

ballooning spider

BALLOONING spiderlings climb onto fence posts or branches and release silk. As the line lengthens, the wind lifts the little spider off its perch and floats it off to a new area. The masses of ballooning threads seen on fall days are called gossamer.

DRAGLINES of silk are laid down by most spiders. Fastened at intervals, they may serve as safety lines or to retrace a path.

SNARES of web spiders are a unique use of silk for trapping insects. Each kind has its special method of catching potential victims. Among them are cobwebs, sheet, funnel, and orb webs. Spiral silk in orbs is sticky in some, woolly in others (p. 106).

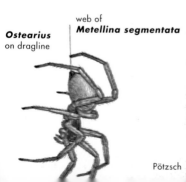

Ostearius on dragline

web of *Metellina segmentata*

Pötzsch

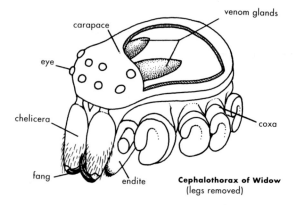

Cephalothorax of Widow
(legs removed)

Labels: venom glands, carapace, eye, chelicera, fang, endite, coxa

POISONOUS SPIDERS, specifically those dangerous to man, are few in number. In the United States and Canada, fatalities from wasp and bee stings far outnumber those from spider bites and scorpion stings. Few spiders will bite even when coaxed, and the bites of most of those large enough to penetrate the skin produce no harm at all.

Knowledge about spider venoms is very limited. Often a spider that bites is immediately destroyed or escapes. Even if the bite causes illness, the spider may not be positively identified.

In the U.S., the dangerous spiders include the Widows and the Recluse Spiders. Bites of *Cheiracanthium* (p. 89), small whitish spiders found in most parts of the world, may produce a slight fever and destroy tissues around the bite. Venomous Hobo Spiders (*Tegenaria agrestis*, p. 74) from Europe have been introduced to NW U.S. and Brit. Columbia. Their bites resemble those caused by Recluse Spiders. In Brazil *Lycosa raptoria* (p. 83) and *Phoneutria* (p. 91) are venomous. The Funnelweb Mygalomorph (*Atrax*, p. 24) and a number of other Australian mygalomorphs have dangerous venoms.

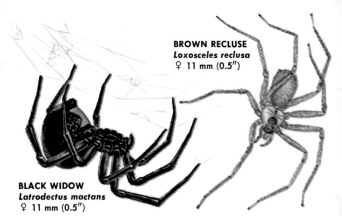

BROWN RECLUSE
Loxosceles reclusa
♀ 11 mm (0.5″)

BLACK WIDOW
Latrodectus mactans
♀ 11 mm (0.5″)

WIDOWS are web spiders. The sedentary females may bite if molested. Males move about but do not bite. Widow spiders (p. 42) are found in most parts of the world. Species occur north to southern Canada and to southern S. A.

The bite may go unnoticed and may not hurt. But the subsequent severe abdominal pain from a Black Widow's bite resembles appendicitis. There is pain also in muscles and in the soles of the feet, but usually no swelling at the site of the bite. Alternately, the saliva flows freely, then the mouth is dry. The bite victim sweats profusely. The eyelids are swollen. The patient usually recovers after several days of agony. Physicians can relieve the severe pain by injection of calcium gluconate. Antivenin is available in all countries where bites occur frequently. No first aid treatment is available for any spider bite.

RECLUSE SPIDERS (*Loxosceles laeta*) of Chile, Peru, and Argentina have been known since the 1930's to cause severe illness. It was not until the 1950's, as a result of bites in Texas, Kansas, Missouri and Oklahoma, that the smaller Brown Recluse spider (*L. reclusa*) was recognized to be similarly toxic. This spider commonly lives in houses on the floor or behind furniture. Bites occur when a spider rests in clothing or in a towel. There may be no harm at all. In very severe cases, a red zone appears around the bite, then a crust forms and falls off. The wound grows deeper and does not heal for several months.

Other species of *Loxosceles* are found in southwestern U.S. and in Mediterranean countries (p. 29). Probably because these spiders do not have contact with man, accidents do not occur. In any bite from a spider known to be poisonous it is wise to consult a physician as soon as signs of illness appear.

17

COLLECTING SPIDERS can be done nearly everywhere—in houses, gardens, fields and woods, under bark, stones, or logs. Turn stones and logs back so the habitat is not destroyed. Spiders that run along the ground can be chased into a glass vial or picked up and dropped into a vial of alcohol. Each collecting vial may be half-filled with specimens. Be sure to insert a penciled field label, citing locality, date, and collector.

SWEEPING shrubs and herbs with an insect net is a technique by which many spiders can be collected. Small spiders can be sifted from leaf litter by using a shaker with cloth sides and a 1 cm (¼") screen bottom. Burrowing spiders must be dug out.

NIGHT COLLECTING with a miner's headlamp yields a harvest of wolf spiders, whose eyes reflect light, and of nocturnal orb weavers that show up against the dark background. Small spiders that hide in crevices by day sit in their webs at night.

A TIN CAN buried flush with the ground surface will trap running spiders. Put in the can a small amount of ethylene glycol (antifreeze), which does not evaporate. A raised lid supported by stones prevents dilution by rain. Empty once a week.

TULLGREN FUNNELS have a 1 cm (¼") wire screen across the bottom. Leaf litter is placed on the screen. Fumes from a single mothball suspended below the lid will drive spiders and other small animals down into a container of water or alcohol below.

headlamp

insect sweep-net

tin can trap

sifter

Tullgren funnel

cotton stoppered vial

jar containing vials

> PERU Junín: Tarma
> 3100m, under bark
> 9 Feb.1965 H.Levi

collecting label

PRESERVATION of spiders and their kin must be in liquid, either 80% grain alcohol or 70–80% iso-propyl (rubbing) alcohol, as these animals are soft-bodied and cannot be pinned and dried. In sorting, keep specimens submerged. Each labeled species should be kept in a separate vial and stored with others in a larger jar of alcohol.

LABELING is most important. Place a label *inside* every vial. Include on the label the date and locality (state, nearest town), collector's name, and habitat. Penciled labels are good in the field. Later replace with labels in ink or typed. Photocopied or laser printed labels do not keep. A specimen without a label is worthless.

REARING Ground spiders can be kept in terraria with soil. Or cut an air hole into a plastic box with a hot knife, and glue screen over it. Web spiders can be kept in wooden frames with glass or cellophane sides. To prevent cannibalism, each spider must be kept in a separate enclosure. Spiders need water, but do not allow containers to get moldy. Many spiders do not need food for days. Spiders will eat living flies, mealworms, cockroaches. Millipedes and wood lice will eat decaying vegetation.

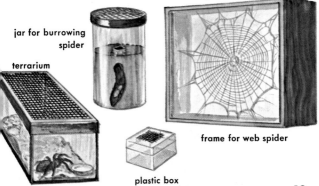

jar for burrowing spider

terrarium

frame for web spider

plastic box

MYGALOMORPHS
Suborder Mygalomorphae

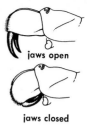

jaws open

jaws closed

Mygalomorphs include the largest spiders. There are about 80 species north of Mexico, many more south; few in Europe. Their jaws (chelicerae), attached on front of the head, move up and down, opening parallel to long axis of body. All have 4 lungs (p. 25).

HAIRY MYGALOMORPHS, TARANTULAS (Theraphosidae). Commonly called tarantulas in the U.S. Unfortunately, this name is shared with other spiders. Hairy Mygalomorphs are known also as Bird Spiders, and they may occasionally catch nestling birds, lizards, or small snakes. In S. Africa, they are called Baboon Spiders. Most are not poisonous to man. About 30 species occur in the U.S., mostly in the Southwest, more in Europe. The largest, from the Amazon Basin of S.A., may be 6–9 cm (3.5") long, with a 25 cm (10") leg span.

Most Hairy Mygalomorphs live on the ground, but some dwell in trees, others burrow. The eyes are closely grouped; these spiders are sensitive to vibrations and hunt at night by touch. Cornered, the spider may purr or rear up on the back legs. The "hairs" on the abdomen, easily shed or rubbed off by the legs, are very irritating to human skin. The underside of each leg tip has a pad of iridescent hairs. Young males look like females, but after the final molt, emerge slender and iridescent, palps developed. Captive females have lived 20 years and molt after maturity; adult males, shorter lived, do not molt.

Members of a related family, Barychelidae (not illustrated), have a digging rake (p. 23) and make a trapdoor to burrow entrance.

Aphonopelma eutylenum
southern California

♂ 60 mm (2.3″)

These spiders show the diversity among the Hairy Mygalomorphs and also their common feature of hairiness.

Cyrtopholis sp.
♀ 50 mm (2″)
Puerto Rico

MEXICAN BLOND TARANTULA

A. chalcodes
♀ 70 mm (2.7″)
Arizona

TRAPDOOR SPIDERS (Ctenizidae) are mainly tropical, but numerous species are found in the southern U.S. and a few in southern Europe. All are about 1-3 cm (0.3-1.2") long. Using the spiny rake on the margins of their jaws, trapdoor spiders dig tubelike burrows. The tube, including the opening, is lined completely with silk. To make the trapdoor, the spider cuts around the rim of the opening, leaving one side attached for the hinge. The top of the lid is camouflaged with debris, and additional silk is added under the lid to make it fit tightly. The lid may be held shut by the spider. When the spider feels the vibration of passing prey, it rushes out, captures the prey, and takes it down the tube. Except to capture prey, the female seldom leaves her tube; males wander in search of mates. Spiders in the small family Migidae (not illustrated), mainly Australian and South African, are similar but lack digging rakes.

CYCLOCOSMIA are found in southeastern U.S. and southeastern China. The spider makes a false bottom for its tube with the hardened, squared-off end of its abdomen and closes the top with a silken lid.

MYRMEKIAPHILA of several species are found in southeastern U.S. The burrow, often located in an ant nest, has a side branch closed by a second door. The outside door to the burrow is covered by a silken lid.

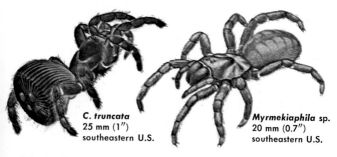

C. truncata
25 mm (1")
southeastern U.S.

Myrmekiaphila sp.
20 mm (0.7")
southeastern U.S.

B. californicum
southern California

trapdoor

25 mm (1")

BOTHRIOCYRTUM found in California, is the most commonly collected trapdoor spider. *Ummidia* (not illustrated) has its third tibia saddle-shaped. Several *Ummidia* are found in the southeastern states, where they dig almost horizontal tubes into banks. The similar *Nemesia* is found in southern Europe.

FOLDING-DOOR SPIDERS, *Antrodiaetus,* occur from the Gulf coast to Alaska. Tube dwellers, they close their tubes by drawing in the rim. Because the anal tubercle is some distance from the spinnerets and there are hardened plates (sclerites) on the back of the abdomen, *Antrodiaetus* is in its own family—Antrodiaetidae.

Antrodiaetus burrow

rake

jaw

Antrodiaetus sp.
15 mm (0.6")
U.S.

FUNNELWEB MYGALOMORPHS (Dipluridae) are easily recognized by their long spinnerets, which may be more than half the length of the abdomen. The spiders are up to 50 mm (2″) in size and most have only four spinnerets. Funnelweb Mygalomorphs catch insects by entangling them in a sheet of silk. The spider hides in a tube in one corner of the sheet. The tube may be among the roots at the base of a tree or in crevices in rocks or wood. Funnelweb Mygalomorphs are mainly tropical, but about ten species are found in the U. S. and a few in Spain. *Atrax* is the poisonous Funnelweb Spider of Australia. The N. A. *Microhexura* is only 3 mm (0.1″) long. Because it has six spinnerets, *Hexura* is sometimes placed in a different family, the Mecicobothriidae.

Hexura fulva
12 mm (0.5″)
U.S. Pacific coast

Euagrus
spinnerets

Hexura
spinnerets

15 mm (0.6″)

Hexura web
and retreat

Euagrus sp.
Texas

PURSEWEB SPIDERS (Atypidae) are about 10 to 30 mm (0.4-1.1″) long. The coxa of each palp is widened to form an endite, and these, as in all true spiders, serve as mouthparts. *Sphodros,* found from Kansas and Texas as far north as Wisconsin and New England, digs a hole at the base of a tree and constructs a silken tube camouflaged with debris. The spider stays hidden inside the tube, which may extend 15 cm (6″) up the side of the tree. If an insect lands on the tube, the spider bites through it with its huge fangs and pulls in the insect. The remains are thrown out through the hole before it is patched up. In the northern states, males are found after June rains when they wander in search of females. The European *Atypus* constructs a small tube that resembles a half-buried root.

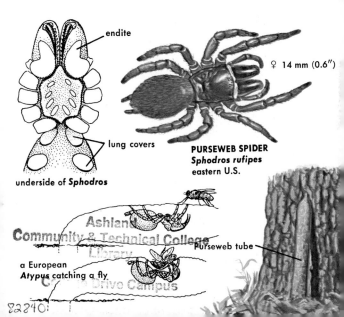

endite

♀ 14 mm (0.6″)

lung covers

PURSEWEB SPIDER
Sphodros rufipes
eastern U.S.

underside of **Sphodros**

Purseweb tube

a European
Atypus catching a fly

TRUE SPIDERS

Suborder Araneomorphae

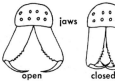

jaws

open · closed

Most common spiders belong to this suborder, found even in the Arctic. Their jaws (chelicerae) are attached below the head and open sideways, sometimes obliquely. With few exceptions, all have two lungs.

carapace

Orchestina saltitans
♀ 1 mm (0.04″)
eastern U.S. buildings

OONOPIDS (Oonopidae) are all less than 3 mm (0.1″) long. Most are short-legged and have six tiny eyes, closely grouped. Many have orange plates on the abdomen. Oonopids live under stones or in litter and can run fast. Most are tropical. About 20 species occur in southern U.S.; several reach northern Europe, some in houses.

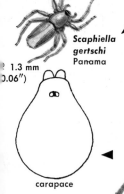

Scaphiella gertschi
Panama
♀ 1.3 mm (0.06″)

Nops sp.
♀ 6 mm (0.2″)
Lesser Antilles

carapace

CAPONIIDS (Caponiidae) to 13 mm (0.5″) long, have only two eyes, rarely eight in one group. The oval abdomen lacks lungs but has four respiratory slits. Found in litter and under stones, in tropics and southwestern U.S.; they run rapidly.

DYSDERIDS (Dysderidae) have six eyes closely grouped, a long labium, and four conspicuous respiratory slits, rather than two, on the underside of the abdomen. All are nocturnal. *Segestria* and *Ariadna*, but not *Dysdera*, direct three pairs of legs forward, one pair back. *Dysdera* lives under stones or bark; it has long jaws, an adaptation for hunting woodlice (p. 152). Others trap insects in silk strands radiating from the opening of a tubular retreat. About 10 species are found north of Mexico, more in Europe.

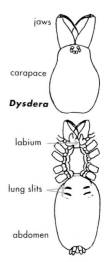

jaws

carapace

Dysdera

labium

lung slits

abdomen

Ariadna bicolor ▼
♀ 12 mm (0.5")
U.S.

Dysdera crocota
▼ ♀ 12 mm (0.5")
N. Hemisphere;
under stones

Ariadna web
with tube extending
into crevice of wall

◄ *Segestria senoculata*
♀ 9 mm (0.4")
Eurasia

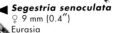

carapace

side view of cephalothorax

SPITTING SPIDERS (Scytodidae) can be recognized by the shape of the carapace. Underneath its dome is a pair of large glands. With their secretion the spider squirts sticky threads at prospective prey and holds it in place. The spider's aim is accurate up to 2 cm (0.7"). Most species are tropical. The female carries her egg sac in her jaws.

Scytodes thoracica
♀ 8 mm (0.3")
cosmopolitan; buildings

carrying eggs

Scytodes fusca
♀ 6 mm (0.2")
cosmotropical

SIX-EYED CRAB SPIDERS (Sicariidae) extend their legs sideways. They live on sand and can dig themselves into it to disappear completely. They are found only in dry regions of S.A. and South Africa.

egg case

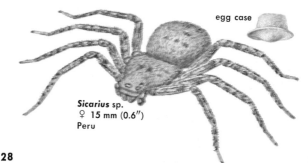

Sicarius sp.
♀ 15 mm (0.6")
Peru

RECLUSE SPIDERS (Loxoscelidae) are six-eyed; their legs do not extend sideways. They weave a sheet of sticky silk in which they entangle insects. *Loxosceles reclusa* (p. 17) in the U.S. and the larger *L. laeta* in S.A. may live in houses with man and are readily transported. Their bite is venomous. Eggs are in a loose sac in the web.

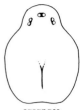

carapace

face

Loxosceles laeta
♀ 25 mm (1")
South America

Loxosceles of several similar species occur in southwestern and south central U.S. *L. rufescens*, of Mediterranean area, lives under stones away from houses.

L. reclusa

other Loxosceles species

L. laeta web

carapace

Diguetia
canities
southwestern
U.S.

DIGUETIDS (Diguetidae) are related to Spitting Spiders (p. 28) and also have six eyes in three groups. The cephalothorax is long, the abdomen hairy. The few species known are from southwestern U.S., Mexico, and Argentina. All members of this family make a vertical silk tube above a maze of threads in desert shrubs.

♀ 8 mm (0.3″)

D. catamarquensis
western Argentina

web, center 60 mm
high (2.1″)

PLECTREURIDS (Plectreuridae) have eight eyes, thick legs, three claws. Like other haplogyne spiders (pp. 8, 26-30), males have simple palps, females no epigynum. Under stones in webs in southwestern U.S., eastern Mexico.

carapace

Plectreurys sp.
12 mm (0.5″)
northwestern Mexico

♀

♂

ZODARIIDS (Zodariidae) are a diverse group of eight-eyed hunting spiders. Altogether there are about 400 species. Unlike haplogyne spiders, male Zodariids have a complicated palpus, and the females have an epigynum. Unlike the Palpimanids, their legs are usually equally thick, and they may have more than two spinnerets. Commonly the first spinnerets are large, those to the rear, small. Zodariids hide under stones or burrow in sand; no species is common in N.A.

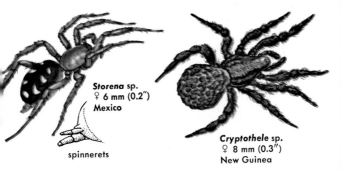

Storena sp.
♀ 6 mm (0.2")
Mexico

spinnerets

Cryptothele sp.
♀ 8 mm (0.3")
New Guinea

PALPIMANIDS (Palpimanidae) are eight-eyed spiders that resemble Zodariids but usually have only two spinnerets. The heavy first pair of legs is carried up when walking. The sternum surrounds first segments (coxae) of legs. They make irregular webs under stones and debris. The family includes about 80 species; none in N.A.

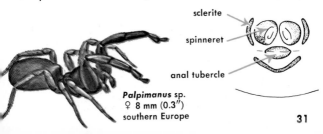

sclerite

spinneret

anal tubercle

Palpimanus sp.
♀ 8 mm (0.3")
southern Europe

31

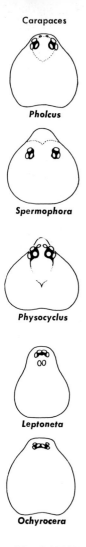

Carapaces

Pholcus

Spermophora

Physocyclus

Leptoneta

Ochyrocera

DADDY-LONG-LEGS SPIDERS (Pholcidae, not to be confused with Phalangiidae, p. 132) have unusually thin, long, slender legs with flexible ends. Most species are whitish or gray. A few have six eyes in two groups of three. Others have eight eyes, the front center pair small. The eyes are always close together. Many of these spiders hang upside down in a loose web in dark corners of houses or cellars. Others live under stones in the dry areas of temperate and subtropical regions. Males and females are commonly found together. Males have large, simple palps; the females lack an epigynum but have a swollen area on the underside of the abdomen. The female carries the round egg sac in her jaws. Of more than 500 species of Pholcids, about 40 species are found in North America north of Mexico; a few occur in northern Europe and many in the Mediterranean region.

There are several other families of rare, small (1–3 mm) (0.05–0.1″), long-legged spiders. These are mostly cave spiders—the Leptonetidae of the Mediterranean region, Japan, and rarely America; and the Ochyroceratidae, which includes about 20 tropical and subtropical species found in America, Africa, and Asia. Carapaces of representative genera are at left.

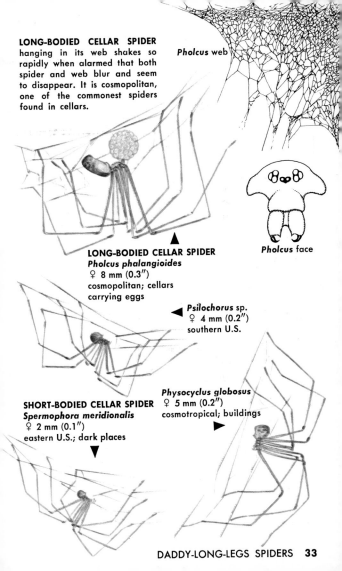

LONG-BODIED CELLAR SPIDER hanging in its web shakes so rapidly when alarmed that both spider and web blur and seem to disappear. It is cosmopolitan, one of the commonest spiders found in cellars.

Pholcus web

Pholcus face

LONG-BODIED CELLAR SPIDER
Pholcus phalangioides
♀ 8 mm (0.3")
cosmopolitan; cellars
carrying eggs

Psilochorus sp.
♀ 4 mm (0.2")
southern U.S.

Physocyclus globosus
♀ 5 mm (0.2")
cosmotropical; buildings

SHORT-BODIED CELLAR SPIDER
Spermophora meridionalis
♀ 2 mm (0.1")
eastern U.S.; dark places

DADDY-LONG-LEGS SPIDERS **33**

UROCTEAS (Oecobiidae) of about a dozen species are found only in the Old World. They live under stones and in rock crevices, where they make a dense, flat silk tube up to 5 cm (2") wide or a series of sheets above and below the spider. Insects crossing the threads become entangled and are spun into the web as the spider, it's spinnerets pointing in, runs around the insect. The prey is then cut loose and carried back to the center of the retreat. The egg sac is also placed between the layers of silk in the retreat. Large urocteids are oecobiids (p. 115) but lack the transverse plate, or cribellum, in front of the spinnerets, with which oecobiids spin their distinctive webs.

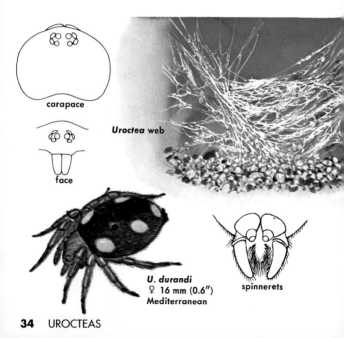

carapace

face

Uroctea web

U. durandi
♀ 16 mm (0.6")
Mediterranean

spinnerets

HERSILIIDS (Hersiliidae) form a family of 100 tropical and subtropical species. They are 10–18 mm (0.4–0.7″) long, with distinctively long spinnerets. Hersiliids position themselves head-down on bark or stone walls. When an insect approaches, the spider jumps over it, spreading silk and then rapidly circles, spinnerets toward the prey, fastening it down. After the prey is completely wrapped, it is bitten and eaten. Members of only one genus, *Tama*, are found in southern Florida, Texas.

carapace

face

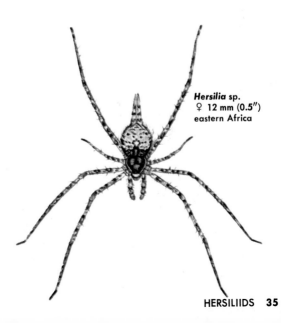

Hersilia sp.
♀ 12 mm (0.5″)
eastern Africa

COBWEB WEAVERS, or Combfooted Spiders (Theridi-idae), with more than 2,100 species, make up one of the large families of common spiders. More than 230 species occur in N.A. north of Mexico, fewer in Europe; many species are tropical and cosmopolitan. The American House Spider (p. 40) and the Widows (p. 42) are members of this family.

Cobweb Weavers are usually sedentary, hanging upside-down in the center of an irregular cobweb or hiding in a crevice at the edge of the web. Some make a small web beneath leaves, stones, or loose bark. The sticky outside threads entangle an insect that hits them and may pull it into the web as they contract. Using a tiny comb of bristles (setae) at the end of the fourth leg, the spider throws silk over the captive, then bites and sucks it dry. Cobweb Weavers have few or no teeth and do not chew prey, as do spiders of related groups.

Most spiders in this family lack strong setae (hairs) on legs. Many have a spherical abdomen, almost all have eight eyes, and all have three claws on each leg (p. 13). As in other web spinners, the male has poor vision and courts the female by plucking threads of her web.

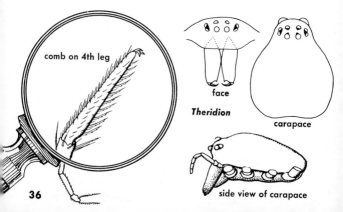

comb on 4th leg

face

Theridion

carapace

side view of carapace

36

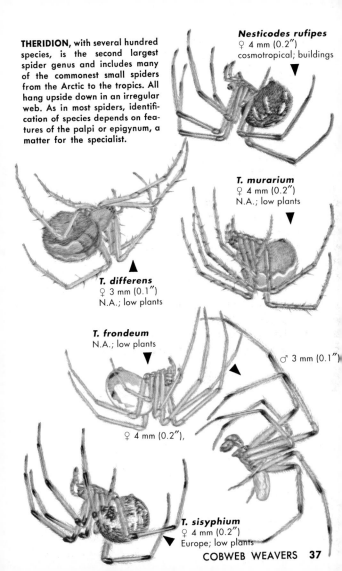

THERIDION, with several hundred species, is the second largest spider genus and includes many of the commonest small spiders from the Arctic to the tropics. All hang upside down in an irregular web. As in most spiders, identification of species depends on features of the palpi or epigynum, a matter for the specialist.

Nesticodes rufipes
♀ 4 mm (0.2″)
cosmotropical; buildings
▼

T. murarium
♀ 4 mm (0.2″)
N.A.; low plants
▼

T. differens
♀ 3 mm (0.1″)
N.A.; low plants
▲

T. frondeum
N.A.; low plants
▼

♂ 3 mm (0.1″)

♀ 4 mm (0.2″),

T. sisyphium
♀ 4 mm (0.2″)
Europe; low plants
▼

COBWEB WEAVERS **37**

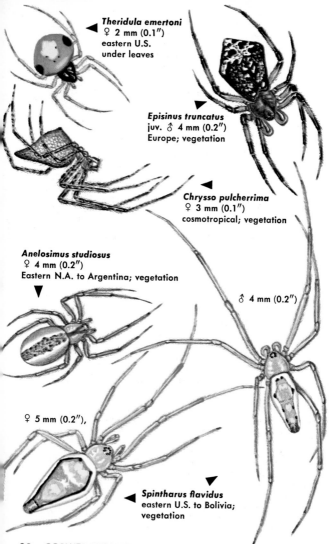

Theridula emertoni
♀ 2 mm (0.1")
eastern U.S.
under leaves

Episinus truncatus
juv. ♂ 4 mm (0.2")
Europe; vegetation

Chrysso pulcherrima
♀ 3 mm (0.1")
cosmotropical; vegetation

Anelosimus studiosus
♀ 4 mm (0.2")
Eastern N.A. to Argentina; vegetation

♂ 4 mm (0.2")

♀ 5 mm (0.2"),

Spintharus flavidus
eastern U.S. to Bolivia;
vegetation

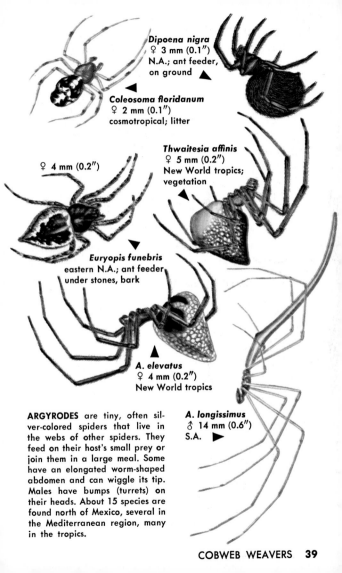

Dipoena nigra
♀ 3 mm (0.1")
N.A.; ant feeder,
on ground

Coleosoma floridanum
♀ 2 mm (0.1")
cosmotropical; litter

♀ 4 mm (0.2")

Thwaitesia affinis
♀ 5 mm (0.2")
New World tropics;
vegetation

Euryopis funebris
eastern N.A.; ant feeder
under stones, bark

A. elevatus
♀ 4 mm (0.2")
New World tropics

A. longissimus
♂ 14 mm (0.6")
S.A. ▶

ARGYRODES are tiny, often silver-colored spiders that live in the webs of other spiders. They feed on their host's small prey or join them in a large meal. Some have an elongated worm-shaped abdomen and can wiggle its tip. Males have bumps (turrets) on their heads. About 15 species are found north of Mexico, several in the Mediterranean region, many in the tropics.

COBWEB WEAVERS **39**

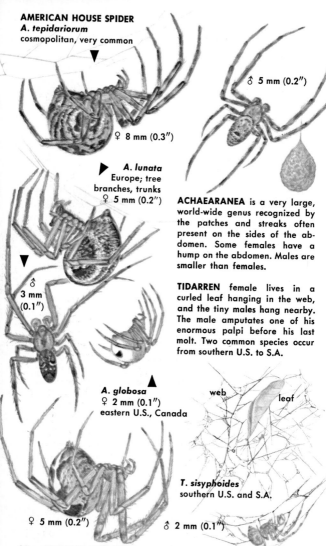

AMERICAN HOUSE SPIDER
A. tepidariorum
cosmopolitan, very common

♂ 5 mm (0.2″)

♀ 8 mm (0.3″)

A. lunata
Europe; tree
branches, trunks
♀ 5 mm (0.2″)

ACHAEARANEA is a very large, world-wide genus recognized by the patches and streaks often present on the sides of the abdomen. Some females have a hump on the abdomen. Males are smaller than females.

TIDARREN female lives in a curled leaf hanging in the web, and the tiny males hang nearby. The male amputates one of his enormous palpi before his last molt. Two common species occur from southern U.S. to S.A.

♂ 3 mm (0.1″)

A. globosa
♀ 2 mm (0.1″)
eastern U.S., Canada

web   leaf

T. sisyphoides
southern U.S. and S.A.

♀ 5 mm (0.2″)   ♂ 2 mm (0.1″)

40 COBWEB WEAVERS

♀ 7 mm (0.3")

underside

E. tecta
northeastern U.S.; plants ▼

♀ 5 mm (0.2")

E. ovata ▼
Europe, northeastern and
northwestern N.A.; plants

ENOPLOGNATHA are all dark colored (except *E. ovata)* and have a leaflike pattern on the abdomen. Some live in curled-up leaves, some under logs, others in leaf litter. Several species occur in N.A. and Europe.

STEATODA usually are dark brown with a white line around the front of the abdomen. *S. borealis* sits in a crevice near the web. *S. hespera,* of western N.A., and *S. bipunctata,* of Europe, are similar.

S. borealis
♀ 8 mm (0.3") ▲
eastern N.A.; tree trunks,
buildings

▼ *S. triangulosa*
♀ 5 mm (0.2")
cosmopolitan

◄ *S. grossa*
♀ 9 mm (0.4")
cosmopolitan

S. erigoniformis
♀ 3 mm (0.1")
cosmotropical; under stones
▼

WIDOWS *(Latrodectus)* are the best known and largest of the Cobweb Weavers. Several species are found in the U.S., one in southern Europe, and the others in the Near East and S.A. All are venomous (pp. 16–17). Females are about 12–16 mm (0.5–0.6") long; males much smaller and with longer legs. Adult males wander in search of females but do not feed or bite; females rarely leave their web. Strands of silk are very strong.

BLACK WIDOW (L. mactans) is found in warm southeastern U.S. and West Indies as far North as New York. It is common in trash, outhouses, and dumps. The spider hangs in the web, which is under objects. In North America many spider bites are from Black Widows. Similar species are found in the western states, Mexico, and other parts of the world. The western North American species is L. hesperus. Both L. mactans and L. hesperus have an hourglass mark on the underside. The abdomen of the Malmignatte (L. tredecimguttatus), the northern Mediterranean species, is marked with a series of red spots. The young of all species are brightly colored with stripes and spots. The egg sac is brown and papery.

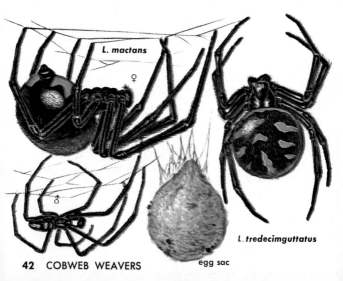

L. mactans ♀

♂

L. tredecimguttatus

egg sac

BROWN WIDOW *(L. geometricus)* cosmotropical, introduced in Florida. Usually they are brown to gray, some black; they are found on or near buildings. Though poisonous it is less likely to bite or injects less venom than other Widows. The egg sac is tufted, fluffy.

egg sac

♀

RED WIDOW *(L. bishopi)* is found only in palmettos in the sandy scrub-pine of central and southern Florida. The egg sac is white, smooth.

♀

NORTHERN WIDOW *(L. variolus)* occurs from northern Florida to southern Canada; common in British Columbia. Generally it is found in undisturbed woods, in dumps, or in stone walls. The egg sac is brown, paper-like. The "hourglass" is distinct but usually broken.

♀

L. variolus
L. bishopi

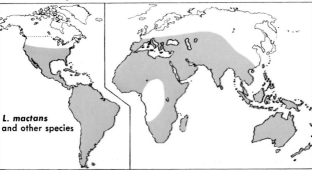

L. mactans and other species

DWARF SPIDERS AND SHEET-WEB WEAVERS (Linyphiidae) form one of the largest families of spiders, with more than 3,600 species. They are among the least known of all spiders, however. Like other web spinners, they have three claws on each foot, and most species build a dome-shaped or flat web. All have legs with strong setae, a row of teeth on each side of the fang groove of the jaws, and a colulus (p. 9).

DWARF SPIDERS (subfamily Micryphantinae) of probably over 600 species occur in N.A. north of Mexico. They are common in Europe, the Arctic, and on high mountains; the tropics are believed to have fewer species. Most Dwarf Spiders are less than 2 mm (0.1") long, some less than 1 mm (0.05") long. Most make small sheet webs. The males of many Dwarf Spiders have unusual turrets, bulges, or depressions in the head region. The abdomen is usually spherical and, in some, it is covered with a hard, shiny plate.

Some species are abundant in leaf litter and can be collected in large numbers with a Tullgren funnel (p. 18). Many can be found under stones; others can be collected by sweeping vegetation with a net. As many as 11,000 spiders per acre (0.4 hectare) have been found in eastern U.S., over 2¼ million per acre (6 million per hectare) in a grassy area in England. These figures include all the different species of spiders, but the Dwarf Spiders make up two-thirds of the total.

The great number of insects consumed daily by such large populations of little predators is difficult to estimate. How will this natural food chain be affected by the spraying of insecticides over the countryside? It will favor the insects, because the spiders lay fewer eggs and each generation takes longer to mature.

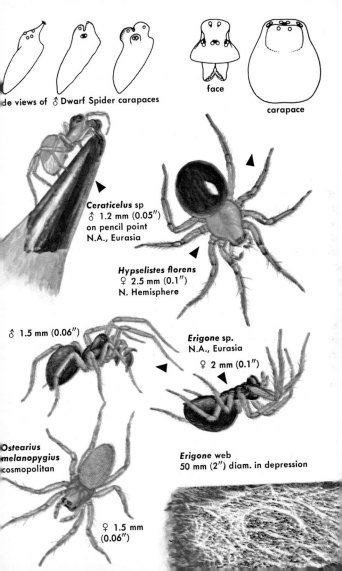

de views of ♂ Dwarf Spider carapaces

face

carapace

Ceraticelus sp
♂ 1.2 mm (0.05")
on pencil point
N.A., Eurasia

Hypselistes florens
♀ 2.5 mm (0.1")
N. Hemisphere

♂ 1.5 mm (0.06")

Erigone sp.
N.A., Eurasia

♀ 2 mm (0.1")

*Ostearius
melanopygius*
cosmopolitan

♀ 1.5 mm
(0.06")

Erigone web
50 mm (2") diam. in depression

SHEETWEB WEAVERS (subfamily Linyphiinae) are mostly larger than Dwarf Spiders (p. 44) and usually have a pattern on the abdomen. In contrast to Cobweb Weavers, the abdomen is usually longer than wide, the legs may have strong setae, the jaws many teeth. The male and female often hang upside down under the same web and run rapidly when disturbed. The webs, which have a few sticky threads, are found between branches of trees or bushes and in high grass, often in great abundance. If an insect gets entangled, the spider bites from below, pulls the insect through the sheet and wraps it up. The web also protects the spider from predators from above, and sometimes a second web forms protection from below. Only a few of the many species are widespread and abundant.

♂ 5 mm (0.2″)

face

carapace

HAMMOCK SPIDER
Pityohyphantes sp.
N. Hemisphere; trees

♀ 7 mm (0.3″)

web 20 cm (8″)

web 10 cm (4")

PLATFORM SPIDER
Microlinyphia sp.
♀ 5 mm (0.2")
N. Hemisphere;
high grass

FILMY DOME SPIDER
Prolinyphia marginata
♀ 5 mm (0.2")
N. Hemisphere; shrubs

web 15 cm (6")

web 15 cm (6")

BOWL AND DOILY SPIDER
Frontinella communis
eastern and central
Canada and U.S.

♀ 4 mm (0.2")

SHEETWEB WEAVERS **47**

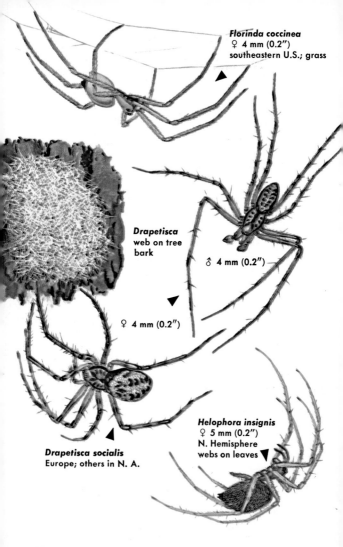

Florinda coccinea
♀ 4 mm (0.2″)
southeastern U.S.; grass

Drapetisca
web on tree
bark

♂ 4 mm (0.2″)

♀ 4 mm (0.2″)

Helophora insignis
♀ 5 mm (0.2″)
N. Hemisphere
webs on leaves

Drapetisca socialis
Europe; others in N. A.

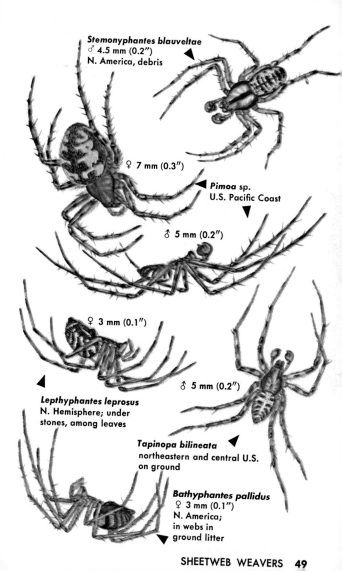

Stemonyphantes blauveltae
♂ 4.5 mm (0.2")
N. America, debris

♀ 7 mm (0.3")

Pimoa sp.
U.S. Pacific Coast

♂ 5 mm (0.2")

♀ 3 mm (0.1")

♂ 5 mm (0.2")

Lepthyphantes leprosus
N. Hemisphere; under
stones, among leaves

Tapinopa bilineata
northeastern and central U.S.
on ground

Bathyphantes pallidus
♀ 3 mm (0.1")
N. America;
in webs in
ground litter

SHEETWEB WEAVERS **49**

PIRATE SPIDERS (Mimetidae) invade webs of other spiders. The slow-moving Pirate Spider bites the web owner, which is quickly paralyzed and sucked dry through the legs, one after another. Some sit with outstretched legs under leaves waiting for passing spiders. One species has been observed to pluck the prey's web like a courting male to gain entrance.

Pirate Spiders are recognized by the row of strong curved setae on the front margins of the lower segments of the first pair of legs. The eggs are left in a stalked sac suspended from a twig or rock. This small family includes about a dozen species north of Mexico.

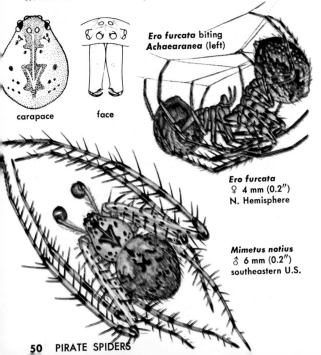

carapace

face

Ero furcata biting
Achaearanea (left)

Ero furcata
♀ 4 mm (0.2")
N. Hemisphere

Mimetus notius
♂ 6 mm (0.2")
southeastern U.S.

CAVE WEAVERS (Nesticidae), often called Cave Spiders, are pale denizens of moist caves and cellars. They make an irregular cobweb. The fourth leg has a comb on its last segment (p. 36). Between the front pair of spinnerets is a colulus (p. 9). This structure, its function not known, is found also in some Cobweb Weavers, but all of these are dark spiders. The Female Nesticid spider carries her egg sac attached to her spinnerets.

Thirty-one species of Nesticids occur north of Mexico. Relatives include the minute Symphytognathidae (not illustrated), the minute Archaeidae (*Archaea*, below) of Africa, and Mecysmauchenidae of South America.

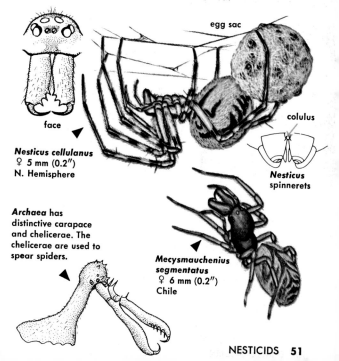

egg sac

face

colulus

Nesticus cellulanus
♀ 5 mm (0.2")
N. Hemisphere

Nesticus
spinnerets

Archaea has distinctive carapace and chelicerae. The chelicerae are used to spear spiders.

Mecysmauchenius segmentatus
♀ 6 mm (0.2")
Chile

NESTICIDS **51**

ORB-WEAVERS (Araneidae, often called Argiopidae) form a family of some 3,000 species found in all parts of the world. About 160 species occur north of Mexico. Almost all of these spiders spin an orb web, an engineering feat practiced also by the Uloborids (p. 114) and by the related Ray Spiders and Tetragnathids (pp. 70–71).

Orb-weavers have poor vision. They locate prey by feeling the vibration and tension of the threads in their web, then quickly turn the captive with their legs while their fourth legs pull out silk and wrap the victim. The prey is bitten before being carried to the center of the web or to the spider's retreat in a corner, where it is eaten. Anything inedible is cut out of the web and dropped to the ground.

In the fall, female Orb-weavers of many species produce egg sacs containing several hundred eggs (p. 14), then die. In some species the eggs hatch soon; in others not until the following spring. The large number of eggs produced suggests that these spiderlings face greater hazards than do the young of scorpions, pseudoscorpions, and spiders cared for by their mothers. Orb-weaver spiderlings make a perfect orb web, but as the spiderlings mature, their webs become more specialized and characteristic of the species.

Orb webs are a favorite object of research in instinctive behavior. Strands are pulled back while the spider is working to learn how the spider compensates for the change. Or, as the spider builds a web in a frame, the frame is turned to determine the influence of gravity on the position of the web. Changes in the web-building pattern as the spider matures are studied, too. Small doses of some drugs given to the spider on a fly will also cause the spider to change the pattern of its web.

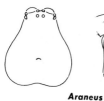

Araneus
carapace

Araneus
face

Argiope
face

Argiope
carapace

SILVER ARGIOPE
Argiope argentata

grasshopper caught in web
is rapidly turned with first
legs while fourth legs wrap
it in silk sheet drawn
from spinnerets

Prey-handling
By Typical
Orb-Weavers

♀ 16 mm (0.6")

Araneus quadratus

carries wrapped prey; will
hang it up or feed on it.

♀ 13 mm (0.5")

ORB-WEAVERS **53**

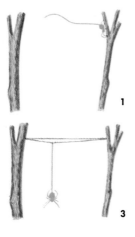

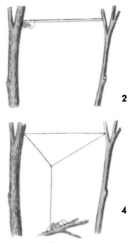

ORB WEBS are built by many species at night. First a bridge is made. The spider sits with its abdomen in the air and lets the wind pull out a silk thread (1). If the thread touches, a bridge is established (2); if not, it may be pulled back and eaten.

When a bridge is established, the spider may reinforce it by walking back and forth, laying down more silk. Then the spider drops on a thread (3) it has fastened at the center of a strand in the bridge. It secures the vertical thread and returns to the fork, the hub of the final web (4).

A radius thread, attached at the hub, is carried up to the bridge and across a short distance (5) before it is tightened and fastened (6). More radii are formed by the same procedure (7–9), and the hub may be strengthened with additional threads before the spider starts a temporary spiral (10). Once around, a leg touches the previous turn, thus measuring the distance between rounds. When the temporary spiral is completed, the spider reverses direction, rolls up the old and puts down new, more numerous, and closer spaced spirals of sticky silk (11). The complete spiral path back to the hub is retraced (12).

Most rebuild the radii and spirals each day or night, some remove the web during the day. A central decoration characteristic of the species may be added, or a silken retreat in rolled up leaves may be made at one side of the web, with a direct line to the hub that transmits vibrations of insects caught in the web. There are many variations from this simplified description. Dusting lightly with corn starch will make the silk visible.

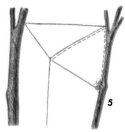

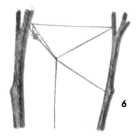

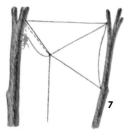

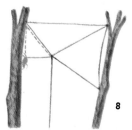

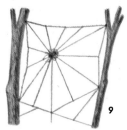

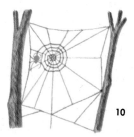

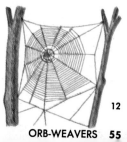

web 50 cm (20")

ARANEUS is the largest genus of spiders, with over 1,500 species found in most parts of the world. Many species make a retreat in a rolled-up leaf near the web. Web variations are transmitted to the retreat from a signal line attached to the hub.

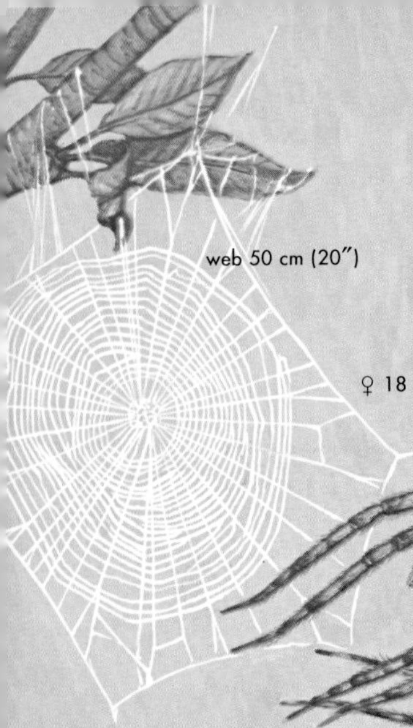

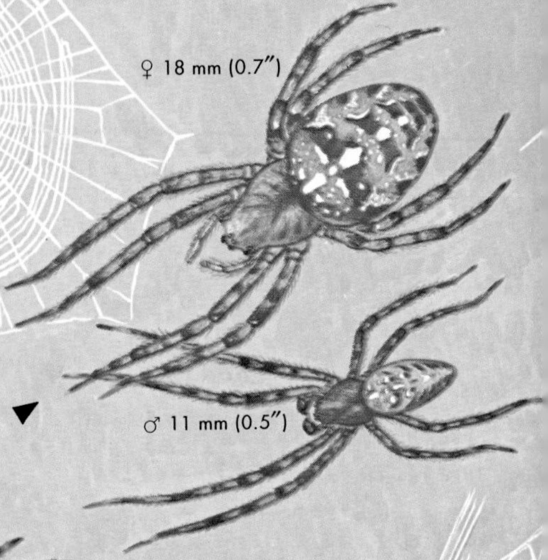

♀ 18 mm (0.7")

♂ 11 mm (0.5")

ROSS ORB-WEAVER ▼
. diadematus
. Hemisphere

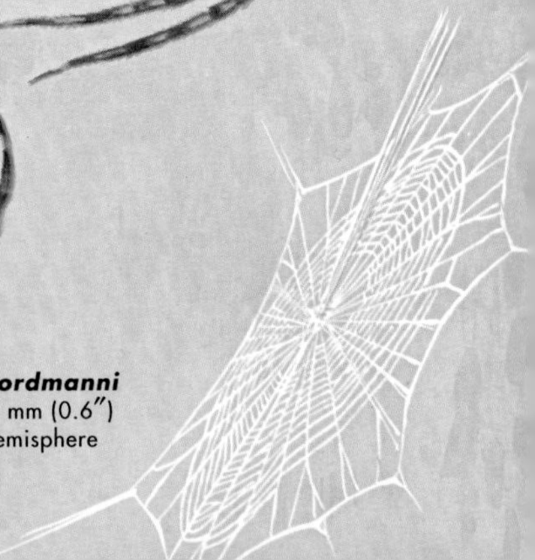

A. nordmanni
♀ 15 mm (0.6")
N. Hemisphere

web 70 cm (28")

♀ 18 mm
(0.7")

♂ 6 mm (0.2")

SHAMROCK ORB-WEAVER
A. trifolium
widespread in N.A.; meadows
♀ may be white

MARBLED ORB-WEAVER
A. marmoreus
♀ 15 mm (0.6")
widespread in N. Hemisphere;
meadows

A. quadratus
♀ 14 mm (0.6")
Eurasia; moist
meadows

A. saevus
♀ 20 mm (0.8")
N. Hemisphere

♀ 20 mm (0.8″) ♂ 15 mm (0.6″)

▲
BARN ORB-WEAVER
Araneus cavaticus
eastern U.S. and Canada

underside

♂ 9 mm (0.4″)

♀ 12 mm (0.5″)

▲
FURROW ORB-WEAVER
Larinioides cornutus
widespread in
N. Hemisphere; on buildings

◄ *Larinioides patagiatus*
♀ 11 mm (0.5″)
N. Hemisphere

LATTICE SPIDER►
Araneus thaddeus
♀ 8 mm (0.3″)
eastern U.S.

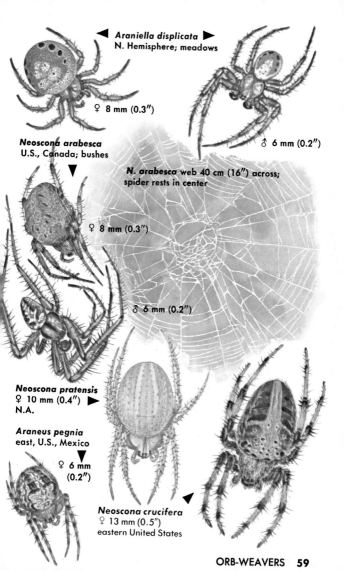

◀ *Araniella displicata* ▶
N. Hemisphere; meadows

♀ 8 mm (0.3")

♂ 6 mm (0.2")

Neoscona arabesca
U.S., Canada; bushes
▼

N. arabesca web 40 cm (16") across;
spider rests in center

♀ 8 mm (0.3")

♂ 6 mm (0.2")

Neoscona pratensis
♀ 10 mm (0.4") ▶
N.A.

Araneus pegnia
east, U.S., Mexico

♀ 6 mm
(0.2")
▼

Neoscona crucifera
♀ 13 mm (0.5")
eastern United States
◀

ORB-WEAVERS 59

META, METELLINA, now placed in Tetragnathidae (p. 70), have theridiid-like abdomen (p. 37). *Meta* is found in dark cellars, cave entrances, wells.

web, 25–40 cm (10–16"); usually inclined with spider in center

♀ 10 mm (0.4") with egg s

CAVE ORB-WEAVER
Meta ovalis
eastern U.S., similar species in Eurasia

♂ 7 mm (0.3")

♀ 4 mm (0.2")

♀ 8 mm (0.3")
Metellina merianae
Europe

Metellina curtisi
western U.S.

60 ORB-WEAVERS

Zygiella x-notata
♀ 8 mm (0.3")
Europe, U.S. Pacific
and Atlantic coasts

Z. x-notata web

Araneus pratensis
♀ 5 mm (0.2")
U.S., Canada ▶

Acacesia hamata
♀ 6 mm (0.2")
eastern U.S. ◀

Zilla diodia
juv. 3 mm (0.1")
Europe ▼

▲
Zygiella atrica
♀ 8 mm (0.3")
Europe, Brit. Columbia
and eastern Canada to New England

ORB-WEAVERS **61**

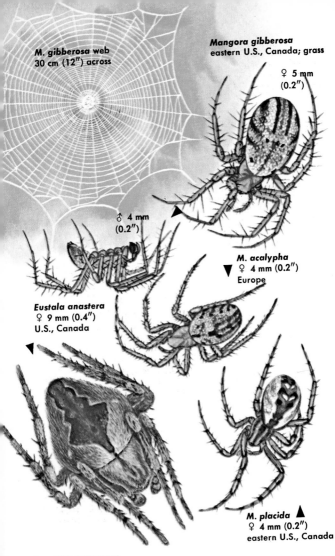

M. gibberosa web
30 cm (12") across

Mangora gibberosa
eastern U.S., Canada; grass

♀ 5 mm
(0.2")

♂ 4 mm
(0.2")

M. acalypha
♀ 4 mm (0.2")
Europe

Eustala anastera
♀ 9 mm (0.4")
U.S., Canada

M. placida
♀ 4 mm (0.2")
eastern U.S., Canada

CYCLOSA hang in the hub, hidden by debris in a vertical line. Egg sac is in center of web. The spider's abdomen extends beyond spinnerets.

C. conica
♀ 5 mm (0.2")
N. A., Europe;
woodlands

♀ 6 mm (0.2")

C. turbinata
N.A.

♂ 3 mm (0.1")

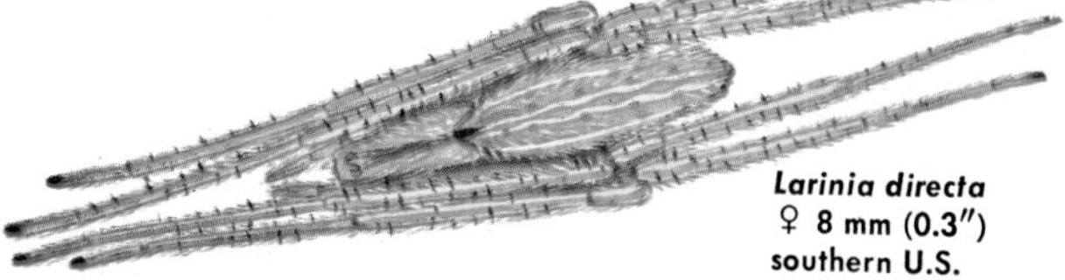

Larinia directa
♀ 8 mm (0.3")
southern U.S.

Mecynogea web
50 cm (20") across
Argentina

Metepeira web
vertical orb-web, with
cobweb and retreat on side

BASILICA SPIDER
Mecynogea lemniscata
♀ 13 mm (0.5")
southeastern U.S.;
bushes

Metepeira sp.
♀ 8 mm (0.3")
New World

Cyrtophora citricola
♀ 15 mm (0.6")
Mediterranean;
has web like *Mecynogea*

Aculepeira sp.
♀ 20 mm (0.8")
western U.S., mountain
meadows; *A. ceropegia*
widespread in Eurasia

GOLDEN SILK ORB-WEAVER
N. clavipes
New World tropics

♀ 25 mm (1")

♂ 4 mm (0.2")

NEPHILA, found in southern U.S. and in the tropics, makes a huge web, 1 m (39") or more in diameter. The strong webs are used by South Sea Islanders for bags and fish nets. Young *Nephila* make a complete web; adults build only the bottom portion, leaving the top irregular. *Nephila* has conspicuous tufts of hair on the legs. Females vary in size. Now placed in Tetragnathidae (p. 70).

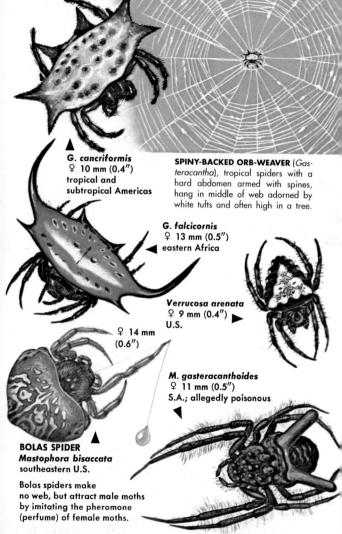

G. cancriformis
♀ 10 mm (0.4")
tropical and
subtropical Americas

SPINY-BACKED ORB-WEAVER (*Gasteracantha*), tropical spiders with a hard abdomen armed with spines, hang in middle of web adorned by white tufts and often high in a tree.

G. falcicornis
♀ 13 mm (0.5")
eastern Africa

Verrucosa arenata
♀ 9 mm (0.4")
U.S.

♀ 14 mm
(0.6")

M. gasteracanthoides
♀ 11 mm (0.5")
S.A.; allegedly poisonous

BOLAS SPIDER
Mastophora bisaccata
southeastern U.S.

Bolas spiders make
no web, but attract male moths
by imitating the pheromone
(perfume) of female moths.

Micrathena sp. in web; southeastern Brazil

ARROWSHAPED MICRATHENA
M. sagittata
♀ 8 mm (0.3″)
eastern U.S.; gardens

MICRATHENA includes spiders with a spiny, hard, glossy abdomen. Several N.A. species are found in woods and gardens; many occur in American tropics.

♀ 10 mm (0.4″)

WHITE MICRATHENA
M. mitrata
♀ 5 mm (0.2″)
eastern U.S.; woods

♂ 5 mm (0.2″)

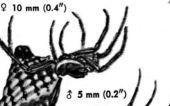

SPINED MICRATHENA
M. gracilis
eastern U.S.; woods

STAR-BELLIED SPIDER
Acanthepeira stellata
♀ 12 mm (0.5″)
eastern U.S.; low shrubs

▼

Isoxya sp.
♀ 8 mm (0.3″)
eastern Africa

ORB-WEAVERS 67

young Black and Yellow
Argiope in web

**BLACK AND YELLOW
ARGIOPE**
A. aurantia
♀ 25 mm (1″)
U.S., Canada; gardens

ARGIOPES are large, conspicuous
spiders that hang head down in
center of web. The web usually
has crossed zigzag bands, and
the young spiders may construct
more zigzags than the adults.
Some species are easily recognized by their color and pattern.
Species of *Argiope* are found in
tropics and temperate regions.

BANDED ARGIOPE
A. trifasciata
♀ 25 mm (1″)
cosmopolitan; fields

BRUENNICH'S ARGIOPE
A. bruennichi
♀ 25 mm (1″)
Europe

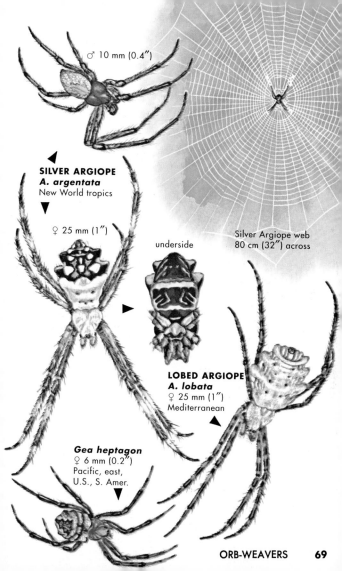

♂ 10 mm (0.4")

SILVER ARGIOPE
A. argentata
New World tropics

♀ 25 mm (1")

underside

Silver Argiope web
80 cm (32") across

LOBED ARGIOPE
A. lobata
♀ 25 mm (1")
Mediterranean

Gea heptagon
♀ 6 mm (0.2")
Pacific, east,
U.S., S. Amer.

ORB-WEAVERS   69

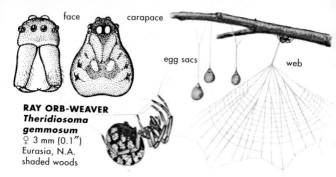

RAY ORB-WEAVER
Theridiosoma gemmosum
♀ 3 mm (0.1″)
Eurasia, N.A.
shaded woods

RAY ORB-WEAVERS (Theridiosomatidae) form a small family of tiny spiders related to the other Orb-weavers (p. 52). The small web, only 10 cm (4″) in diameter, lacks a hub but has several radii tied together near the center. The spider holds up the web in the center by a tight thread so that it forms an umbrella. If a fly gets caught, the thread is released, causing the web to spring back and entangle the catch. The spider has a globular abdomen, the sternum is short and square behind. The egg sac is suspended on a stalk. One species is widespread; about 70 are known from the tropics.

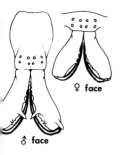

♀ face

♂ face

Tetragnatha sp.

LONG-JAWED ORB-WEAVERS (Tetragnathidae) make an orb web, usually at an angle between vertical and horizontal. The orb usually has 12 to 20 radii and widely spaced spirals. The spider hangs in the center or clings to a stalk somewhere near the web. Unlike other Orb-weavers, the female Tetragnathids, except *Leucauge, Meta, Metellina* (p. 60), and *Nephila* (p. 65) lack an epigynum. Thirty-eight species of Tetragnathids occur north of Mexico.

ORCHARD SPIDER
Leucauge sp.
♀ 7 mm (0.3")
Florida

THICKJAWED SPIDER
Pachygnatha sp.
♀ 6 mm (0.2")
eastern U.S.

ORCHARD SPIDERS *(Leucauge)* are common in wooded areas of eastern U.S. The spiders hang in center of horizontal orb. About 170 species are tropical.

THICK-JAWED SPIDERS *(Pachygnatha)* are found under debris or in dense vegetation near water. Young make small orb web on ground; adults make no web.

LONG-JAWED ORB-WEAVERS, *Tetragnatha,* at rest may cling lengthwise along a twig or grass blade, holding on with the short third pair of legs. The long pairs of legs are extended. More than a dozen species are common in meadows near water throughout N.A. and Europe. There are more than 250 species in all parts of the world.

Tetragnatha web

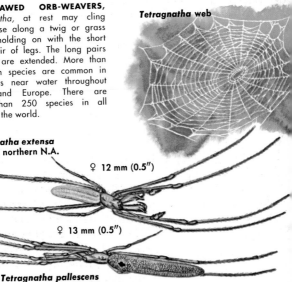

Tetragnatha extensa
Eurasia, northern N.A.

♀ 12 mm (0.5")

♀ 13 mm (0.5")

Tetragnatha pallescens
eastern U.S. to Central America

ORB-WEAVERS 71

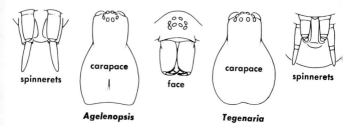

spinnerets

Agelenopsis carapace

face

Tegenaria carapace

spinnerets

FUNNEL WEAVERS (Agelenidae) are seen most easily in late summer when morning dew makes their webs in lawns conspicuous. The spider hides at the narrow end of a funnel that spreads out across the grass. On feeling the vibration of an insect crossing the web, the spider dashes out on the web surface, bites the insect, and carries it back to the funnel. As the spider grows, it uses its long posterior spinnerets to add new layers to the flat web.

Like other web spiders, Funnel Weavers have three leg claws and poor vision. In fall, the female deposits a disc-shaped egg sac in a crevice, then dies—often while still clinging to the egg sac. Of some 500 species in the family, about 300 are found in N.A. and 95 in Europe. Despite lack of a cribellum, the Agelenidae are related to the Amaurobiidae (p. 111).

Agelenopsis web in grass

GRASS SPIDERS (*Agelenopsis* and *Agelena*) make funnel webs in grass or low bushes. There are several similar species of *Agelenopsis* in N.A., each living in a slightly different habitat. *Agelena labyrinthica* is common in Europe.

♂ 15 mm (0.6")

GRASS SPIDER ▶
Agelenopsis sp.
N.A.
◀

♀ 20 mm (0.8")

C. terrestris
♀ 13 mm (0.5")
Europe; leaf litter
▼

COELOTES communicates with her young by making special movements when feeding, summoning them to share the food. As a warning signal, she stamps her fourth leg, and the young scurry into hiding. The mother can distinguish between her young and potential prey by the differences in vibrations in the web and by touch. The young eat the mother when she dies in autumn. Most observations have been made on the European *C. terrestris*.

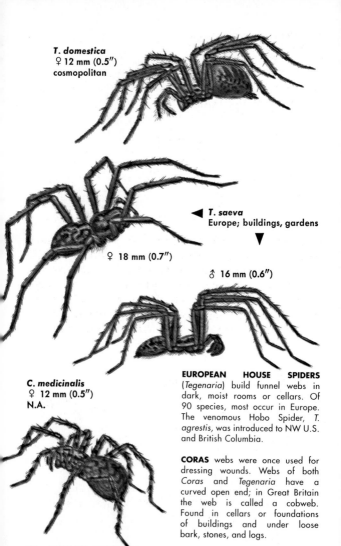

T. domestica
♀ 12 mm (0.5″)
cosmopolitan

◀ **T. saeva**
Europe; buildings, gardens
▼

♀ 18 mm (0.7″)

♂ 16 mm (0.6″)

C. medicinalis
♀ 12 mm (0.5″)
N.A.

EUROPEAN HOUSE SPIDERS (*Tegenaria*) build funnel webs in dark, moist rooms or cellars. Of 90 species, most occur in Europe. The venomous Hobo Spider, *T. agrestis,* was introduced to NW U.S. and British Columbia.

CORAS webs were once used for dressing wounds. Webs of both *Coras* and *Tegenaria* have a curved open end; in Great Britain the web is called a cobweb. Found in cellars or foundations of buildings and under loose bark, stones, and logs.

74 FUNNEL WEAVERS

CICURINA are small funnel weavers that live in leaf litter and under stones. Numerous species in N.A., some in Europe.

CRYPHOECA are found in similar habitats. Species occur in northern U.S., Canada, Europe.

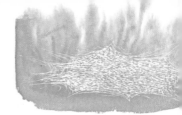

♂ 6 mm (0.2")

♀ 3 mm (0.1")

Cryphoeca sp. northern N.A., Europe; leaf litter, debris

▶ *Cicurina* sp. N.A., Europe; leaf litter, under stones

HAHNIIDS (Hahniidae), shown below, have spinnerets arranged in a single transverse row. Fewer than 225 species are known, 19 north of Mexico. All are small, less than 4 mm (0.2"), and their delicate webs, commonly made in moss or in footprints of animals in moist soil or snow, can be seen only when laden with moisture. The spider lives beneath grains of soil at the edge of the web.

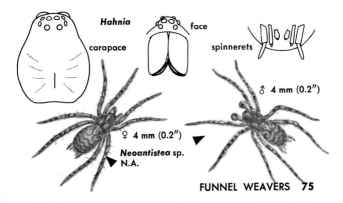

Hahnia
carapace

face

spinnerets

♂ 4 mm (0.2")

♀ 4 mm (0.2")

▶ *Neoantistea* sp. N.A.

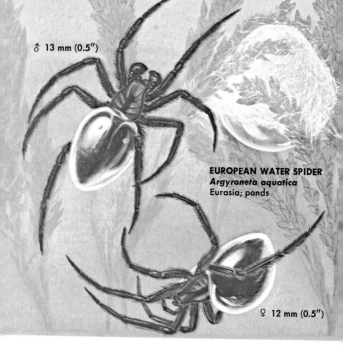

EUROPEAN WATER SPIDER
Argyroneta aquatica
Eurasia; ponds

♂ 13 mm (0.5")

♀ 12 mm (0.5")

EUROPEAN WATER SPIDER (Argyronetidae) is found in ponds, shallow lakes and quiet streams of Europe and Asia. It builds a bell-shaped web among plants under water and fills the bell with air bubbles carried on its body. Occasionally the spider replenishes the air. To do this, the spider comes to the surface, touches it with its first legs, then turns around and projects its abdomen through the surface film. A quick motion with the hind legs replenishes the air carried around the abdomen and under the cephalothorax. The spider swims upside down. It remains dry because of the air clinging to its body.

Aquatic sow bugs and insects are captured and eaten under the bell, and the young are raised there. Unlike most spiders, the males are larger than the females. The bite of these spiders is painful to man.

LYNX SPIDERS (Oxyopidae) are hunting spiders that chase their prey over vegetation or lie in wait and leap out. All are active during daytime and have good vision. Their six large eyes form a hexagon, and there are two smaller eyes below. Lynx Spiders use their silk as a dragline for jumping and for anchoring the egg sac to vegetation, not for catching prey. The female guards the egg sac. Lynx Spiders have three claws on the leg tips. The legs have many long, strong setae. The abdomen is pointed behind. Most of the 400 species are tropical; fewer than 20 species are found north of Mexico and still fewer in Europe.

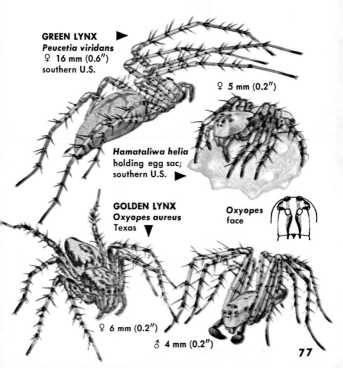

GREEN LYNX ▶
Peucetia viridans
♀ 16 mm (0.6")
southern U.S.

♀ 5 mm (0.2")

Hamataliwa helia
holding egg sac;
southern U.S. ▶

GOLDEN LYNX
Oxyopes aureus
Texas ▼

Oxyopes
face

♀ 6 mm (0.2")

♂ 4 mm (0.2")

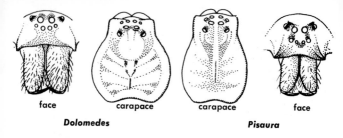

face   carapace   carapace   face

Dolomedes   *Pisaura*

NURSERY WEB SPIDERS (Pisauridae) attract attention because of their large size. They may sit quietly for hours, legs spread out on vegetation or boat docks, or they may hunt actively in vegetation. Their vision is good.

Nursery Web Spiders resemble the related Wolf Spiders (p. 82) but differ in habits. Their eight eyes are about equal in size, and they have three tarsal claws. The female carries her huge egg sac in her jaws. When hatching time is near, she ties leaves together with silk and suspends her egg sac among them. She then sits on guard nearby. The young spiders leave the nursery after about a week.

Many Nursery Web Spiders can run over the surface of water. If chased, they dive and stay submerged for some time. Of about 270 species throughout the world, about 15 occur north of Mexico; fewer occur in Europe.

Pisaura mirabilis
♀ 15 mm (0.6")
carrying egg sac

PISAURINA in N.A. and *Pisaura* of Europe are common spiders. They have no permanent homes but hunt in grass, meadows, and moist, open woods. When courting, the male *Pisaura* presents the female with a fly. If she accepts, he will mate while she feeds.

Pisaura mirabilis ▶
♀ 15 mm (0.6")
Europe

◀ *Pisaurina mira*
6 mm (0.6")
ern U.S., Canada

Pisaurina mira
nursery in milkweed

NURSERY WEB SPIDERS **79**

RAFT SPIDER
D. fimbriatus
♀ 20 mm (0.8")
Eurasia

FISHING SPIDERS are called Raft Spiders in England due to the erroneous belief that they construct rafts. All are large and are found in or near water. They capture insects and, rarely, tadpoles or small fish. There are about 10 species north of Mexico. Most belong to the genus *Dolomedes*.

♀ 20 mm (0.8")

SIX-SPOTTED FISHING SPIDER
D. triton
U.S., E. of Rocky Mts.

♂ 10 mm (0.4")

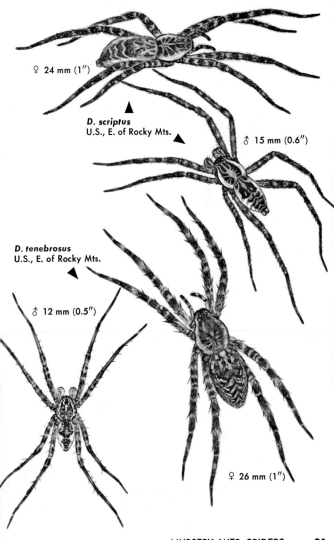

♀ 24 mm (1″)

D. scriptus
U.S., E. of Rocky Mts.

♂ 15 mm (0.6″)

D. tenebrosus
U.S., E. of Rocky Mts.

♂ 12 mm (0.5″)

♀ 26 mm (1″)

WOLF SPIDERS (Lycosidae) are among the most common spiders. They run on the ground or over stones, and some may venture up plants. At rest they stay under stones. Some dig short tunnels, others deep burrows. Many hunt during the day or, in warm climates, at night. They have good vision and a highly developed sense of touch. Males wave their large, often hairy pedipalps in a rhythmic pattern as they approach potential mates.

The female attaches her large egg sac to her spinnerets. If the egg sac is removed, the spider searches, and upon finding the sac attaches it to her spinnerets again. She may substitute bits of cork, paper, or snail shells for the egg sac if lost. As the young spiderlings emerge, they climb onto their mother, who carries them on her back, brushing them away from her eyes. If any fall off, they climb up the mother's legs again. Wolf Spiders make nice pets. In captivity they must be provided with water.

About 2,200 species of Wolf Spiders are known. Perhaps more than 230 occur north of Mexico. Some species are widespread over the Northern Hemisphere; others are very local. They make up a large proportion of the spider population in the Arctic and on high mountains. Wolf Spiders have four small eyes in a row below four larger eyes, and three tarsal claws.

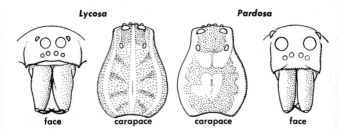

Lycosa

face   carapace

Pardosa

carapace   face

Hogna frondicola
♀ 14 mm (0.6")
with egg sac
northeastern U.S.

GLADICOSA, HOGNA, and **LYCOSA** are large, common Wolf Spiders. Most are nocturnal and can be collected with a light at night. The light is reflected by the eyes. Some hide under objects; others dig burrows, from which they can be lured by inserting a straw or a piece of grass. Included in the genus is the European Tarantula *(L. tarentula)* of Mediterranean countries. Its bite was alleged to be poisonous, its victims cured only by dancing the tarantella. The bites may have been confused with those of Widows (p. 42), as the European Tarantula is not now considered venomous.

Lycosa lenta ▲
♀ 25 mm (1")
carrying young
southeastern U.S.

Gladicosa gulosa
♂ 14 mm (0.6")
U.S., Canada

Lycosa tarentula
♂ 25 mm (1"),
♀ (not illus.) browner and larger
southern Europe
▼

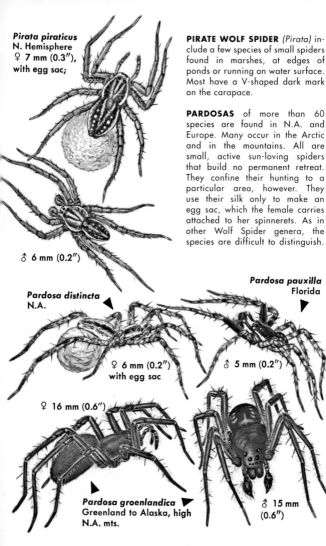

Pirata piraticus
N. Hemisphere
♀ 7 mm (0.3"),
with egg sac;

♂ 6 mm (0.2")

PIRATE WOLF SPIDER *(Pirata)* include a few species of small spiders found in marshes, at edges of ponds or running on water surface. Most have a V-shaped dark mark on the carapace.

PARDOSAS of more than 60 species are found in N.A. and Europe. Many occur in the Arctic and in the mountains. All are small, active sun-loving spiders that build no permanent retreat. They confine their hunting to a particular area, however. They use their silk only to make an egg sac, which the female carries attached to her spinnerets. As in other Wolf Spider genera, the species are difficult to distinguish.

Pardosa pauxilla
Florida

Pardosa distincta
N.A.

♀ 6 mm (0.2")
with egg sac

♂ 5 mm (0.2")

♀ 16 mm (0.6")

Pardosa groenlandica ▶
Greenland to Alaska, high
N.A. mts.

♂ 15 mm
(0.6")

ARCTOSAS are not as common as other wolf spiders, but the species are widespread throughout the Northern Hemisphere. Their color may match the background on which they run.

BURROWING WOLF SPIDERS (*Geolycosa*) dig in sand, as deep as 1 m (39") straight down. They dig with their chelicerae, and the sand grains are stuck together with silk and thrown out of the hole. The different colors of sand from the various levels may form concentric rings of varied colors around the hole. The spider spends most of its time in the burrow; at night it is close to the surface. In fair weather the egg sac is brought up and sunned.

Arctosa sanctarosae
♀ 12 mm (0.5")
U.S. Gulf Coast; sand

♀ 14 mm (0.6")

A. littoralis
U.S., Canada; sand

Geolycosa in entrance

eolycosa burrow

♀ 23 mm (0.9")

Sosippus floridanus
Florida; makes funnel web ◄

♀ 13 mm (0.5")

Burrowing Wolf Spider ▲
Geolycosa missouriensis eastern U.S.; sand

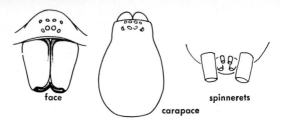

face

carapace

spinnerets

GROUND SPIDERS (Gnaphosidae) are usually uniformly black, a few brown, some with markings. The long abdomen is slightly flattened, and the front spinnerets are cylindrical and separated. The legs have only two claws (p. 13). All are nocturnal hunters and in daytime are found under stones or loose bark, often in a silken sac. The egg sac may be a shiny pink or white papery disc attached tightly to the underside of a stone; in some species it is a white sac guarded by the female. Most Gnaphosidae have posterior median eyes oval at an angle (see above) and the endites (p. 9) concave and slightly constricted in their middle. Many of the 2,100 species of Gnaphosids are found in temperate regions, probably 250 species in N.A. and fewer in Europe. *Micaria* (p. 89) are ground spiders.

Gnaphosa muscorum
N. Hemisphere

Zelotes fratris
N. Hemisphere

♀ 14 mm
(0.6")

♀ 9 mm
(0.4")

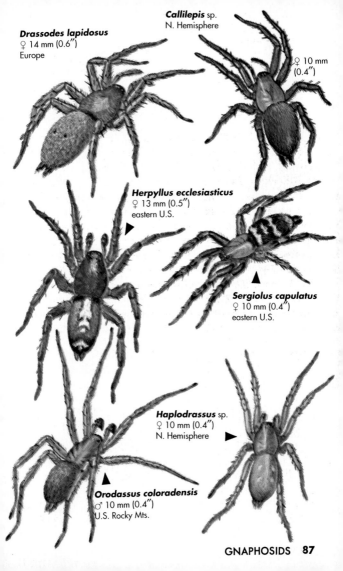

Drassodes lapidosus
♀ 14 mm (0.6″)
Europe

Callilepis sp.
N. Hemisphere

♀ 10 mm
(0.4″)

Herpyllus ecclesiasticus
♀ 13 mm (0.5″)
eastern U.S.

Sergiolus capulatus
♀ 10 mm (0.4″)
eastern U.S.

Haplodrassus sp.
♀ 10 mm (0.4″)
N. Hemisphere

Orodassus coloradensis
♂ 10 mm (0.4″)
U.S. Rocky Mts.

GNAPHOSIDS **87**

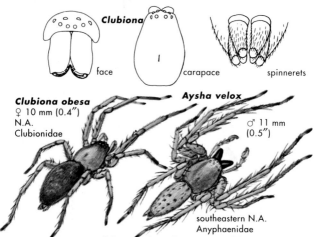

Clubiona

face | carapace | spinnerets

Clubiona obesa
♀ 10 mm (0.4″)
N.A.
Clubionidae

Aysha velox
♂ 11 mm
(0.5″)

southeastern N.A.
Anyphaenidae

SAC SPIDERS all have 2 leg claws, resemble Gnaphosids with abdomen less flattened and front spinnerets more conical. Anyphaenidae (483 species; 37 in U.S. & Canada) have respiratory tube openings at middle of abdomen underside. Corinnidae (660; 37) have anterior spinnerets adjacent. Clubionidae (590; 60) have labium longer than wide, eye region wider than half of carapace. Liocranidae (245; 70) have oblique, depressed endites, labium length equal to or less than width, and paired central spines on anterior tibiae and tarsi.

C. subsaltans ♀ 8 mm (0.3″), with young; Eurasia   photo Pötzsch

Clubiona abboti
♂ 5 mm (0.2")
N.A.

Cheiracanthium mildei
juv. 10 mm (0.4")
Mediterranean; N.A. in
buildings. First leg longer
than last; many species venomous
Clubionidae

ANT MIMICS are abundant in the genera *Micaria* and *Castianeira*. They often live with the ants they mimic, but the advantage of mimicry to the spider is not understood. The abdomen may be constricted or covered with scales; the gait is antlike, and the first legs are held like antennae. *Micaria*, mainly a north temperate genus, prefers dry areas. *Castianeira*, mainly of New World tropics, has a grooved thorax and usually is larger and more brightly colored than *Micaria*.

Phrurotimpus borealis
northeastern N.A.
litter
Liocranidae

♀ 3 mm
(0.1")

Micaria longipes
♀ 5 mm (0.2")
U.S., E. of Rocky Mts.
Gnaphosidae

Castianeira floridana
♂ 8 mm (0.3")
northern Florida, Cuba
Corinnidae

SAC SPIDERS 89

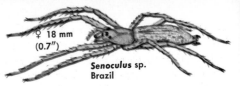

♀ 18 mm
(0.7")

Senoculus sp.
Brazil

Senoculid
carapace

SENOCULIDS (Senoculidae) include only about 31 species in the American tropics. They have three claws and are related to Nursery Web Spiders (p. 78) but differ in arrangement of eyes. Senoculids hunt on plants. The female guards her egg sac.

Prodidomid
carapace

face

PRODIDOMIDS (Prodidomidae) include 60 species with two claws. Related to Gnaphosids (p. 86) but differ in eye arrangement and in having long, spread chelicerae. They are found under stones in dry areas in southern N.A. and in southern Europe.

HOMALONYCHIDS (Homalonychidae), 2 species are found only in Mexico and in southwestern U.S. The legs of these spiders have two claws, may be held straight.

Homalonychid
carapace

face

♀ 14 mm
(0.6")

Homalonychus sp., northwestern Mexico

WANDERING SPIDERS (Ctenidae), 470 species, 5–40 mm (0.3–1.5″) tropical, have anterior lateral eyes in line with posterior medians. Undersides of legs have strong setae. They hunt on vegetation. *Acanthoctenus* (p. 112) has a cribellum. Similar *Zora* (Zoridae) (p. 14), temperate, hunts like wolf spiders.

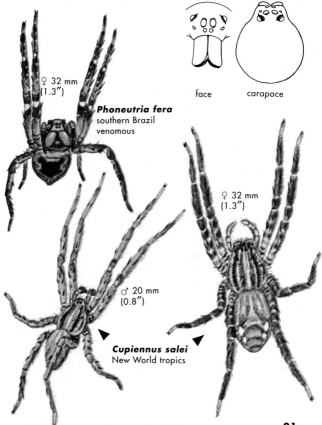

♀ 32 mm
(1.3″)

Phoneutria fera
southern Brazil
venomous

face   carapace

♀ 32 mm
(1.3″)

♂ 20 mm
(0.8″)

Cupiennus salei
New World tropics

GIANT CRAB SPIDERS (Sparassidae or Heteropodidae) are mostly tropical spiders that hold their two-clawed legs crablike. Larger than Crab Spiders (p. 94), they have teeth on their jaws. About 850 species.

HUNTSMAN SPIDER *(Heteropoda venatoria),* found around the world in tropical regions, is welcomed in houses because it eats cockroaches. The spiders hide in crevices during day and come out in evening. The females carry their egg sacs with the jaws. They are found in the southern U.S. and are commonly imported with bananas. A similar huntsman, *Olios,* occurs in the southwestern states and in southern Europe. The front middle eyes of *Olios* are as large or larger than the lateral eyes.

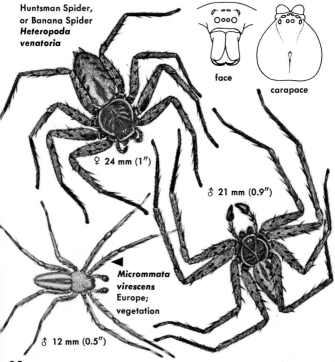

Huntsman Spider, or Banana Spider
Heteropoda venatoria

face

carapace

♀ 24 mm (1″)

♂ 21 mm (0.9″)

► *Micrommata virescens* Europe; vegetation

♂ 12 mm (0.5″)

92

PLATORID CRAB SPIDERS (Platoridae) of about a dozen species occur in Asia and in tropical America. They are characterized by flatness, long middle spinnerets, widely spaced front spinnerets, and two leg claws.

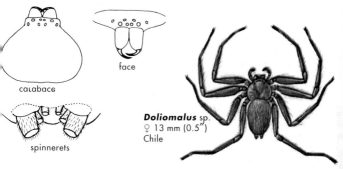

caLabace

face

spinnerets

Doliomalus sp.
♀ 13 mm (0.5″)
Chile

SELENOPID CRAB SPIDERS (Selenopidae) include 160 species of large tropical spiders easily recognized by their flatness and by their eye arrangement—six in a single row. Like other crab spiders, they have two leg claws. Common in houses and under bark or rocks; if disturbed, they dash sideways into crevices.

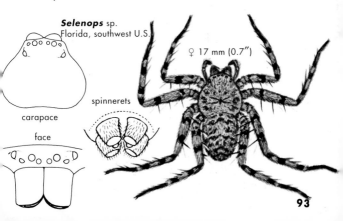

Selenops sp.
Florida, southwest U.S.

♀ 17 mm (0.7″)

spinnerets

carapace

face

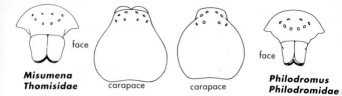

Misumena
Thomisidae

face

carapace

carapace

face

Philodromus
Philodromidae

CRAB SPIDERS (Thomisidae, Philodromidae) hold their legs crablike, out at the sides, and can walk forward, backward, or sideways. The largest are tropical species 12–20 mm (0.5–0.8″) in body length. Some Thomisids have ornaments on the head or abdomen, and some mimic bird droppings. Males are smaller than females and have much longer legs. Crab Spiders wait in ambush for passing insects; some hold their front legs outstretched in readiness. Their vision for movements is good. Their jaws are small, and after prey is bitten, it is held above the spider and sucked dry. Those that sit on flowers apparently have a toxin potent to bees, flies, and other insects much larger than themselves. They do not use silk to capture prey. Thomisidae (2,000 species; 128 in U.S. and Canada) have the first and second legs longer and thicker than the third and fourth. Philodromidae (500; 95) have the first and second legs only a little longer than third and fourth.

FLOWER SPIDERS (*Misumena, Misumenops*) sit on flowers, and can change color slowly.

Misumena vatia
♀ 10 mm (0.4″)
N. Hemisphere
biting bee

▲

♂ 4 mm (0.2″)

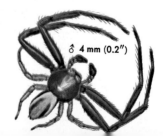

Stephanopis sp.
♀ 6 mm (0.2″)
Chile

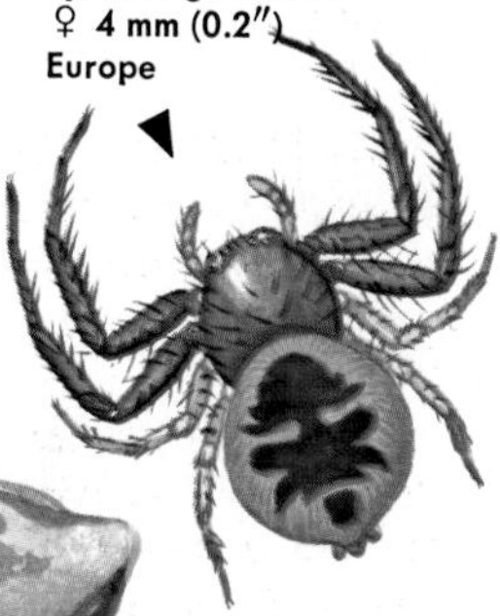

Synema globosum
♀ 4 mm (0.2″)
Europe

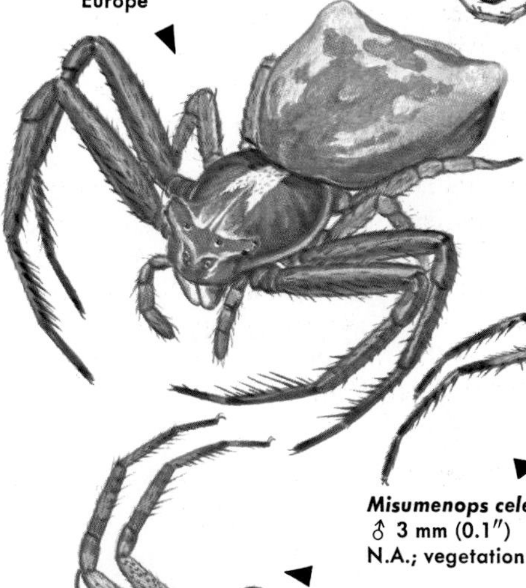

Thomisus onustus
♀ 8 mm (0.3″)
Europe

**THOMISID CRAB
SPIDERS**

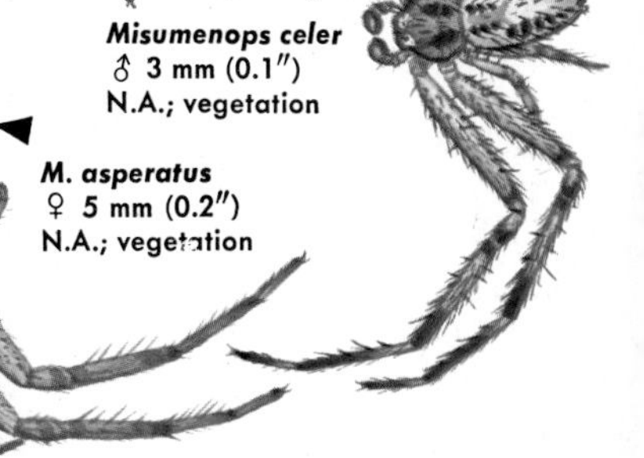

Misumenops celer
♂ 3 mm (0.1″)
N.A.; vegetation

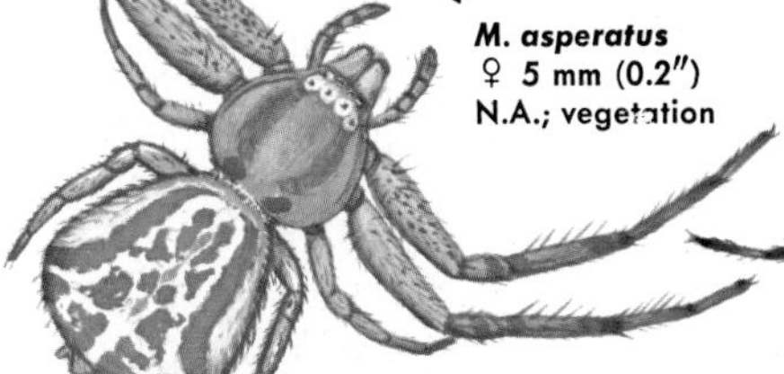

M. asperatus
♀ 5 mm (0.2″)
N.A.; vegetation

X. cristatus
♀ 7 mm (0.3")
Europe

X. emertoni
♂ 5 mm (0.2")
N.A.

THOMISID CRAB SPIDERS

X. alboniger
♀ 5 mm (0.2")
eastern U.S.

XYSTICUS of more than 230 species occur in all parts of the world, mainly in the Northern Hemisphere. Most are found under bark or on the ground. Dull brownish, they resemble their background. Males may look quite different from females.

CORIARACHNE has several species occurring in N.A. Extreme flatness adapts them to hiding in narrow crevices in bark.

♀ 6 mm (0.2")

C. versicolor
eastern N.A.

♂ 5 mm (0.2")

Tmarus angulatus
♀ 7 mm (0.3")
N.A.

Tibellus oblongus
♀ 9 mm (0.4")
N. Hemisphere

Thanatus formicinus
♀ 7 mm (0.3")
N. Hemisphere

PHILODROMID CRAB SPIDERS

TIBELLUS, like long-jawed spiders (p. 71), stretch along twigs and grasses. They usually are collected in grass by sweeping with a net. Males and females are similar in appearance.

THANATUS live on vegetation and bark in N.A. and Europe. A captive *T. formicinus* had a food preference for moths.

PHILODROMUS are active Crab Spiders that climb on bark, plants, or sometimes on ceilings in houses. Egg sacs are fastened to leaves or bark. Many of the more than 170 species occur in N.A. and in Europe.

♀ 8 mm (0.3")

P. praelustris
U.S., Canada

P. rufus
eastern Canada, U.S.,
Europe

♂ 6 mm (0.2")

♀ 4 mm (0.2")

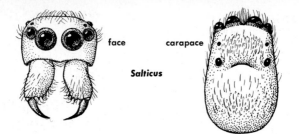

face   carapace

Salticus

JUMPING SPIDERS (Salticidae) are among our most attractive spiders. Most have bright colors, often with iridescent scales. A large family of more than 4,500 species, Jumping Spiders are most abundant in the tropics, but about 300 species occur north of Mexico and many in Europe. All are small, most less than 15 mm (0.7") long.

Jumping Spiders are active during the day and like sunshine. They walk with an irregular gait and leap on their prey, sometimes jumping many times their own length. Most of the jumping power is supplied by the fourth pair of legs, though they are only slightly modified for jumping. Before jumping, the spider secures a silk thread on which it can climb back in case it misses its mark. At night or when it is cool, the spiders stay in little cocoons or in crevices.

Jumping Spiders' eyes are among the best in invertebrate animals. Two of the eight eyes are large. Many species can recognize prey or other spiders at a distance of 10–20 cm (4–8"). They can also change the color of their eyes by moving the retinas.

When a male Jumping Spider finds a female, he stops, waves his brightly colored and enlarged first legs, wags his abdomen, and hops. The third legs may also be colorful and enlarged. If the female is of the same species, she signals with her legs. After mating, the female constructs a silk cocoon for her eggs and guards it.

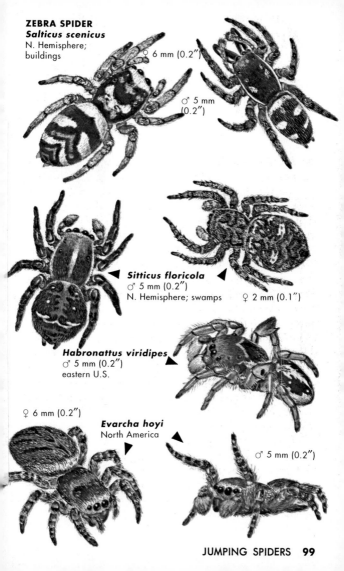

ZEBRA SPIDER
Salticus scenicus
N. Hemisphere;
buildings

♀ 6 mm (0.2")

♂ 5 mm
(0.2")

Sitticus floricola
♂ 5 mm (0.2")
N. Hemisphere; swamps

♀ 2 mm (0.1")

Habronattus viridipes
♂ 5 mm (0.2")
eastern U.S.

♀ 6 mm (0.2")

Evarcha hoyi
North America

♂ 5 mm (0.2")

JUMPING SPIDERS **99**

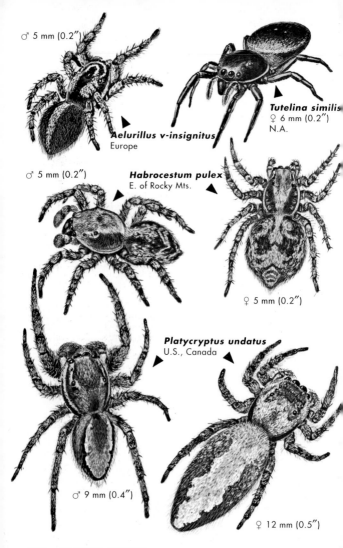

♂ 5 mm (0.2")

Aelurillus v-insignitus
Europe

Tutelina similis
♀ 6 mm (0.2")
N.A.

♂ 5 mm (0.2")

Habrocestum pulex
E. of Rocky Mts.

♀ 5 mm (0.2")

Platycryptus undatus
U.S., Canada

♂ 9 mm (0.4")

♀ 12 mm (0.5")

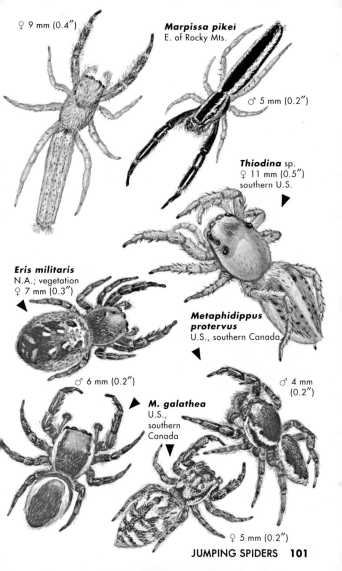

♀ 9 mm (0.4″)

Marpissa pikei
E. of Rocky Mts.

♂ 5 mm (0.2″)

Thiodina sp.
♀ 11 mm (0.5″)
southern U.S.
▶

Eris militaris
N.A.; vegetation
♀ 7 mm (0.3″)
▶

**Metaphidippus
protervus**
U.S., southern Canada
▶

♂ 6 mm (0.2″)

♂ 4 mm
(0.2″)

▶ **M. galathea**
U.S.,
southern
Canada
▼

♀ 5 mm (0.2″)

JUMPING SPIDERS 101

♀ 15 mm (0.6")

♂ 13 mm (0.5")
♂ displaying at ♀

P. audax
E. of Rocky Mts.

PHIDIPPUS is common in N.A.; none in Europe. Large, heavy-bodied and conspicuous, these spiders are found on vegetation, stones, and sometimes inside houses. In captivity, they are active and have good appetites. One captive *Phidippus* ate more than 40 fruit flies in succession.

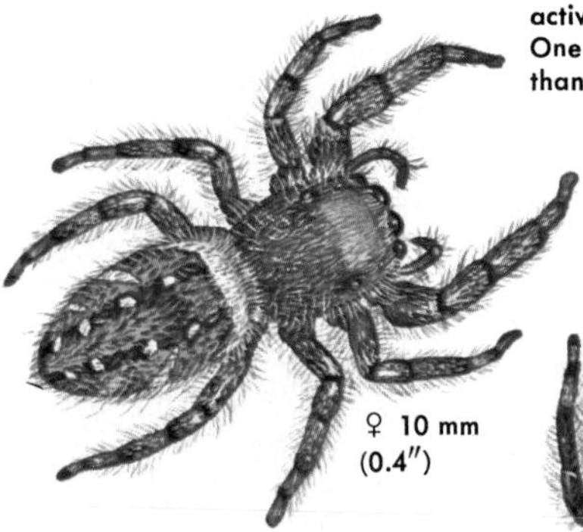

♀ 10 mm (0.4")

♂ 8 mm (0.3")

P. clarus
N.A.; common
on plants

♀ 13 mm (0.5")

♂ 11 mm (0.5")

P. johnsoni
♂ displaying at ♀
Rocky Mts. and westward;
stones

P. regius
♀ 18 mm (0.7")
southeastern U.S.

P. apacheanus
♂ 9 mm (0.4")
U.S.

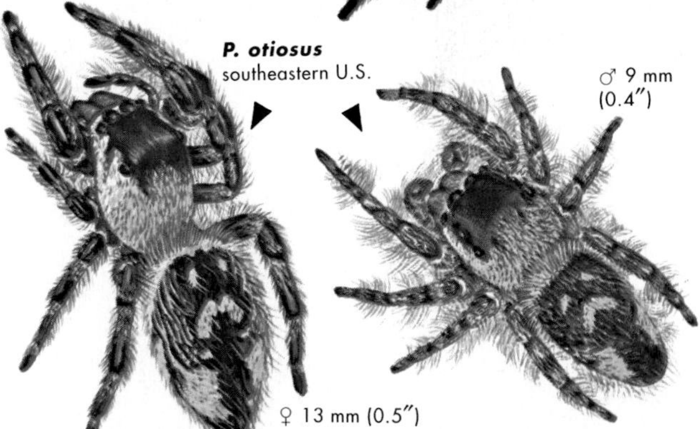

P. otiosus
southeastern U.S.

♂ 9 mm
(0.4")

♀ 13 mm (0.5")

JUMPING SPIDERS **103**

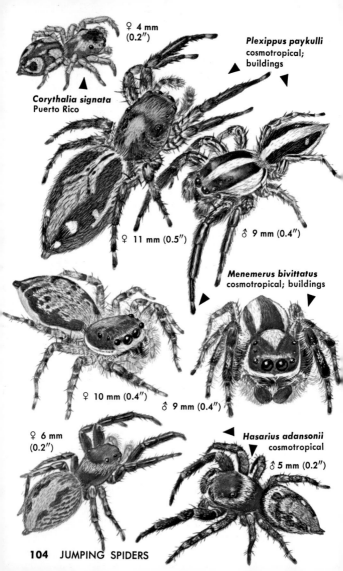

♀ 4 mm
(0.2")

Corythalia signata
Puerto Rico

Plexippus paykulli
cosmotropical;
buildings

♀ 11 mm (0.5")

♂ 9 mm (0.4")

Menemerus bivittatus
cosmotropical; buildings

♀ 10 mm (0.4")

♂ 9 mm (0.4")

♀ 6 mm
(0.2")

Hasarius adansonii
cosmotropical

♂ 5 mm (0.2")

104 JUMPING SPIDERS

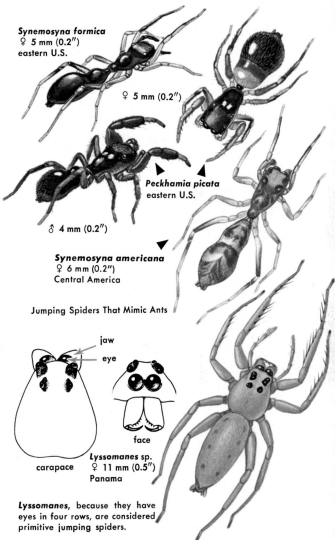

Synemosyna formica
♀ 5 mm (0.2")
eastern U.S.

♀ 5 mm (0.2")

Peckhamia picata
eastern U.S.

♂ 4 mm (0.2")

Synemosyna americana
♀ 6 mm (0.2")
Central America

Jumping Spiders That Mimic Ants

jaw

eye

carapace

face

Lyssomanes sp.
♀ 11 mm (0.5")
Panama

Lyssomanes, because they have eyes in four rows, are considered primitive jumping spiders.

JUMPING SPIDERS **105**

CRIBELLATE SPIDERS have a cribellum, a flattened sievelike plate in front of the spinnerets. Through it the calamistrum, a row of curved setae on the next to last segment of the fourth leg, pulls silk. The hackled threads produced are covered with fine wool that entangles the prey. Every struggling movement of the prey only increases the entanglement.

It was thought that families of cribellate spiders are a separate, distinct group. Now the opinion is that the presence of the cribellum is an ancestral condition and the structure has been lost several times. For instance, Uroctea (p. 34), lacking a cribellum, are placed with the similar but cribellate Oecobiidae (p. 115). The cribellum is still a good character for identification.

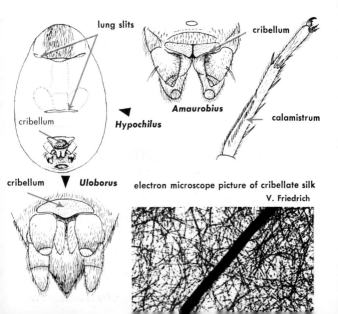

lung slits

cribellum

Amaurobius

Hypochilus

calamistrum

cribellum

cribellum ▼ **Uloborus**

electron microscope picture of cribellate silk
V. Friedrich

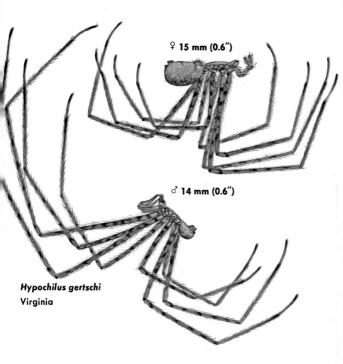

♀ 15 mm (0.6")

♂ 14 mm (0.6")

Hypochilus gertschi
Virginia

LAMPSHADE SPIDERS (Hypochilidae) are found in isolated regions of North America and China. About ten species are included in the family. Some species resemble the Mygalomorphs (p. 20) in jaw attachment and in having two pairs of lungs with lung covers. Under a ledge, *Hypochilus* builds its lampshade-shape web, covered with hackled threads; the spider hangs inside against the ledge. Flies may be caught on the outside of the web and pulled through or they may be trapped under the web canopy.

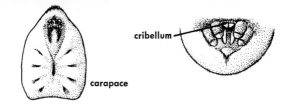

carapace

cribellum

CREVICE WEAVERS (Filistatidae) of fewer than 100 species are common in southern Europe and other warm parts of the world, a dozen in the U.S. The cribellum may be difficult to see, but the shape of the carapace is diagnostic. By day they hide in a tubular retreat in a crevice of a wall. As in *Ariadna* (p. 27), silk strands radiate from the mouth of the tube. Males have much longer legs than females. Females lack an epigynum.

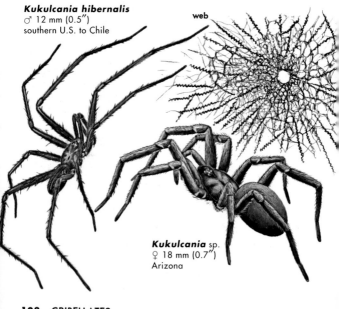

Kukulcania hibernalis
♂ 12 mm (0.5")
southern U.S. to Chile

web

Kukulcania sp.
♀ 18 mm (0.7")
Arizona

ERESIDS (Eresidae) of about 100 species are found in Eurasia and Africa, none in the Americas. Large and hairy, they resemble Mygalomorphs (p. 20). *Eresus* lives under a stone at the end of a silken tube from which a web extends beyond the stone between plant stems. The web resembles that of a Funnel Weaver (p. 72) but is not as clean and new looking. Some have a cover over tube. Stegodyphus lives in shrubs and may be social.

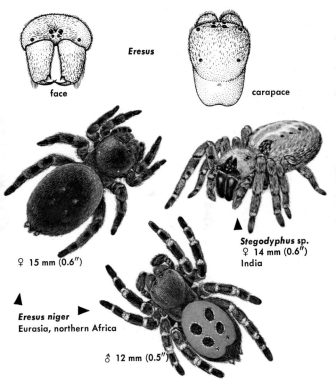

Eresus

face

carapace

Stegodyphus sp.
♀ 14 mm (0.6")
India

♀ 15 mm (0.6")

Eresus niger
Eurasia, northern Africa

♂ 12 mm (0.5")

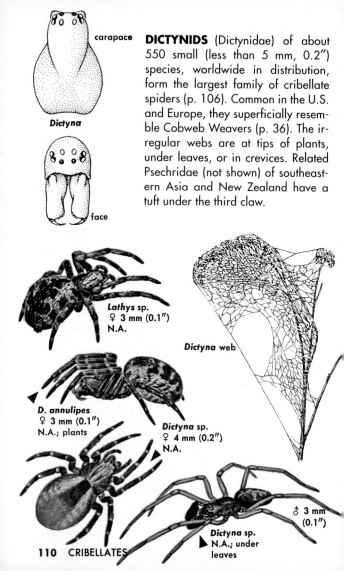

carapace

Dictyna

face

DICTYNIDS (Dictynidae) of about 550 small (less than 5 mm, 0.2") species, worldwide in distribution, form the largest family of cribellate spiders (p. 106). Common in the U.S. and Europe, they superficially resemble Cobweb Weavers (p. 36). The irregular webs are at tips of plants, under leaves, or in crevices. Related Psechridae (not shown) of southeastern Asia and New Zealand have a tuft under the third claw.

Lathys sp.
♀ 3 mm (0.1")
N.A.

Dictyna web

D. annulipes
♀ 3 mm (0.1")
N.A.; plants

Dictyna sp.
♀ 4 mm (0.2")
N.A.

♂ 3 mm (0.1")

Dictyna sp.
N.A.; under leaves

AMAUROBIIDS (Amaurobiidae) are much larger than the Dictynids and are similar to the Funnel Weavers (p. 72). They are found under logs or stones, where they make a loose web with coarse hackling. There are 450 species; the family is worldwide in distribution. Many N.A. and European species are very common; the easily seen divided cribellum distinguishes them from the Funnel Weavers.

carapace

Amaurobius

face

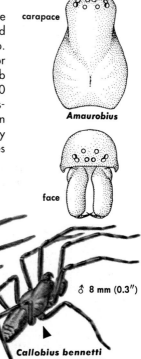

♀ 11 mm (0.5″)

♂ 8 mm (0.3″)

Callobius bennetti
U.S., southern Canada

Amaurobius ferox
eastern N.A.
Europe

♀ 12 mm
(0.5″)

Titanoeca americana
♀ 6 mm (0.2)
eastern U.S.
Canada

CRIBELLATES **111**

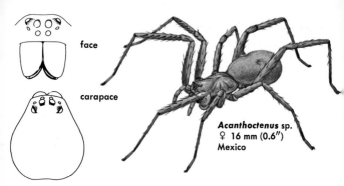

Acanthoctenus sp.
♀ 16 mm (0.6″)
Mexico

ACANTHOCTENIDS (Ctenidae) are Wandering Spiders (p. 91) but have a cribellum. They have two leg claws. Their hackled webs are found near their hiding places under loose bark. About 25 species are known from the American tropics.

ZOROPSIDS (Zoropsidae) have two or three leg claws; most cribellates have three. They differ from Gnaphosids (p. 86) by the cribellum and by scopulae (p. 13) under the last two leg segments. Two dozen species occur around the Mediterranean, in Mexico, and in Central America.

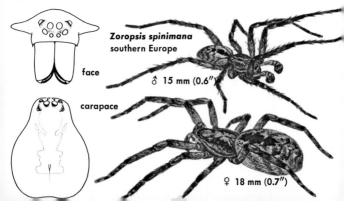

Zoropsis spinimana
southern Europe

♂ 15 mm (0.6″)

♀ 18 mm (0.7″)

OGRE-FACED SPIDERS (Deinopidae) usually have two enormous posterior median eyes; the remaining six eyes are small. During daytime they hide in shrubs, and at night, with a few parallel hackled threads, the spider makes a rectangular web supported by silk lines, which it holds with its front legs. When an insect approaches, the web is spread and thrown. Cosmotropical in distribution, about 50 species occur in the tropics, one in southeastern United States. *Menneus,* of Australia, has small posterior median eyes.

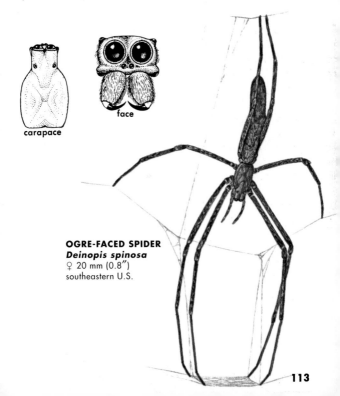

carapace

face

OGRE-FACED SPIDER
Deinopis spinosa
♀ 20 mm (0.8″)
southeastern U.S.

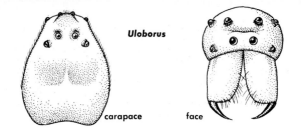

Uloborus

carapace   face

HACKLED BAND ORB-WEAVERS (Uloboridae) have a cribellum but make orb webs, though the spiders are not similar in appearance to Orb-weavers (p. 52). Uloborids lack venom glands. About 250 species occur in all parts of the world. Fifteen species are found in N.A., fewer in Europe, but they are most abundant in the tropics.

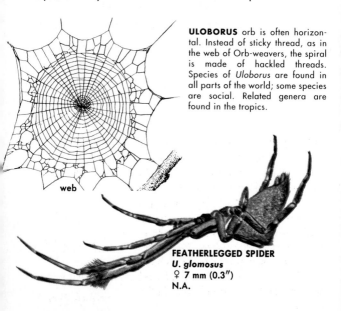

ULOBORUS orb is often horizontal. Instead of sticky thread, as in the web of Orb-weavers, the spiral is made of hackled threads. Species of *Uloborus* are found in all parts of the world; some species are social. Related genera are found in the tropics.

web

FEATHERLEGGED SPIDER
U. glomosus
♀ 7 mm (0.3")
N.A.

HYPTIOTES web is triangular. The spider attaches itself to a twig by silk from its spinnerets and holds the web with its first legs. The spider itself bridges a gap in the threads. When an insect gets caught, the spider pulls the web taut, then lets it go slack again. This further entangles the prey. The web is made by constructing a bridge, a vertical line, and two more radii. Scaffolding threads run out from the center, and hackled threads are put down separately in each sector. *H. cavatus* is found in eastern U.S., southern Canada. *H. paradoxus* is the common European species.

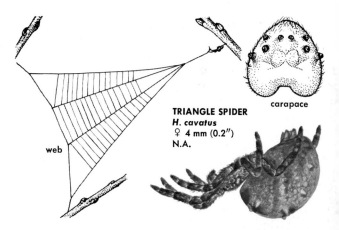

web

carapace

TRIANGLE SPIDER
H. cavatus
♀ 4 mm (0.2")
N.A.

OECOBIIDS (Oecobiidae) include the small *Oecobius* with cribellum and the large *Uroctea* of the Old World, without cribellum (p. 34). About 90 species. Tiny oecobiids make small flat webs over crevices in walls and on various leaves. They feed on ants, and some are social. They have a large, hairy anal tubercle. Most species are tropical and subtropical; few inhabit houses in the North.

Oecobius sp.
3 mm (0.1")
N.A.; buildings

A number of groups (orders) are believed to share a common ancestry with spiders (pp. 6-7). All have jaws (chelicerae), pedipalps (sometimes pincerlike), and four pairs of walking legs. They differ in adaptations of the appendages. In some, the abdomen is segmented and joined broadly to cephalothorax.

WHIPSCORPIONS (order Uropygi; about 85 species) occur in some parts of southern U.S., but are most abundant in Central and South America, Asia, and the East Indies. They have no sting but can pinch. All are nocturnal; they have poor vision, but are sensitive to vibrations.

VINEGARONES, or Vinegaroons (Thelyphonidae), aim a vinegar scented mist from a gland at the base of the tail when disturbed. The spray contains acetic acid and a solvent that attacks the exoskeleton of insects. Other tropical species smell of formic acid or of chlorine. Vinegarones burrow in sand or under logs and carry their prey into the burrow. Female carries 20-35 eggs in a membranous sac under her abdomen. The young ride on the mother until their first molt, then become independent.

VINEGARONE
Mastigoproctus giganteus
southern U.S.

largest whipscorpion known; males and females are similar in appearance

8 cm (3")

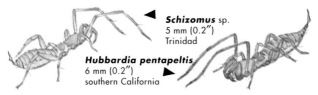

Schizomus sp.
5 mm (0.2")
Trinidad

Hubbardia pentapeltis
6 mm (0.2")
southern California

SHORT-TAILED WHIPSCORPIONS (order Schizomida, sometimes called Schizopeltidia) are small, most less than 6 mm (0.2") long. They lack eyes, and the short tail has 0 to 7 segments. About 180 species live in the tropics and subtropics under stones and in leaf litter.

TAILLESS WHIPSCORPIONS, or Whipspiders (order Amblypygi), have a wide head and thorax, to which the abdomen is attached by a stalk. The first pair of legs is long and whiplike. Amblypygids hide under bark or stones, and if the stone is turned, scurry off sideways. In the U.S., they are found in the warm South, where they may come into houses. Others are found in the tropics. Female carries eggs and young for four to six days. About 80 species are known in five families.

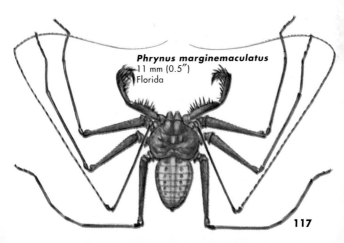

Phrynus marginemaculatus
11 mm (0.5")
Florida

117

WINDSCORPIONS (order Solifugae) are sometimes called Sun Scorpions because most live in deserts or dry areas. From 1 to 5 cm (0.3–2") long, most are yellowish or brown. They have enormous jaws, and their leglike pedipalps are used with the first pair of legs as feelers. They walk on only six legs, as do Whipscorpions (p. 116) and Tailless Whipscorpions (p. 117), and run swiftly—"like the wind." Sensing prey, they may stop suddenly, then start again. Some can climb trees.

Windscorpions are voracious feeders. With their large pincerlike jaws they can kill even small vertebrates and may sometimes feed on small lizards. They can bite but do not have venom glands. It is believed that some use their eyes for hunting, but most use touch only. The mallet-shaped organs on the last pair of legs are probably sensory. At rest, Solifuges may burrow or sit under stones. Most are nocturnal, but some are active during the day. Males are generally smaller and have longer legs than the females. Females bury their eggs and may guard them. Adults live less than a year.

Windscorpions of the Old World belong to several families. They are abundant in Africa and the Near East; absent from Australia. Six species are found in southern Europe. Almost 120 species occur in N.A. north of Mexico, 900 worldwide. They are found as far north as southwestern Canada but are most abundant in Arizona and the Great Basin area. One species occurs in Florida, the only one known from eastern U.S. Windscorpions of N.A. belong to two easily recognized families, but the species are difficult to identify. In the family Ammotrechidae, the first pair of legs have no claws, and the front edge of the head is rounded or pointed. In the family Eremobatidae, the first pair of legs have one or two claws, and the front edge of the head is straight.

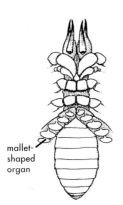

mallet-
shaped
organ

underside of
windscorpion
Solpugidae, eastern Africa

Ammotrechella stimpsoni
20 mm (0.8")
Florida, West Indies

▲

**Eremobates
durangonus**
28 mm (1.1")
Texas to California,
northern Mexico

▲

▲ **Eremobates pallipes**
26 mm (1")
N. Dakota to Arizona

WINDSCORPIONS 119

PSEUDOSCORPIONS (order Pseudoscorpiones), are common everywhere but are rarely seen because of their secretive habits. They are small, less than 5 mm (0.2") long. Of 3,000 species, over 300 are known from the U.S. and Canada. Most species live in leaf litter, moss, manure, under loose bark, or under stones. Many can be collected from litter with a Tullgren funnel (p. 18).

Pseudoscorpions can walk backwards as well as forward. They do not have a long tail and stinger as do scorpions, but most have venom glands in their pincers. The poison is used solely to capture small insects, their prey. Pseudoscorpions are not large enough to bite. They have silk glands opening on their jaws, and use silk to make chambers for overwintering, molting, or brooding. Many species lack eyes and, judging from the long hairs on the pincers, their main sense is touch. Pseudoscorpions sometimes attach themselves to flies or to beetles; they are carried as hitchhikers, not as parasites.

In courting behavior, the male waves his pincers, vibrates his abdomen, or taps his legs. The female responds and the two animals, their pincers locked, pull each other back and forth. Eventually the male deposits on the ground a stalked capsule containing spermatozoa. As the female is pulled over the capsule, she picks it up with the lips of an opening on the underside of her abdomen. The eggs, in a little sac, stay attached to the female's abdomen, and when the young hatch, they remain in the sac and feed on a milklike secretion from the mother's ovaries. Usually there are fewer than two dozen young, but there may be more than one brood a year.

After leaving the egg sac, the young ride on the mother for a short time. It may be one to several years before they become adults; during this period, the young molt three times.

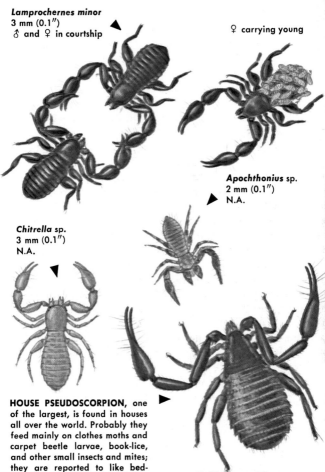

Lamprochernes minor
3 mm (0.1")
♂ and ♀ in courtship

♀ carrying young

Apochthonius sp.
2 mm (0.1")
N.A.

Chitrella sp.
3 mm (0.1")
N.A.

HOUSE PSEUDOSCORPION, one of the largest, is found in houses all over the world. Probably they feed mainly on clothes moths and carpet beetle larvae, book-lice, and other small insects and mites; they are reported to like bed-bugs. In search of moisture they often become stranded in sinks or bath tubs, unable to climb out over the smooth surface.

Chelifer cancroides
4 mm (0.2") cosmopolitan

SCORPIONS (order Scorpiones) have pincers and a long tail with a stinger at its tip. The pectines, a pair of comblike structures underneath the last legs, are sense organs of touch. Scorpions have two eyes in the center of the head and usually two to five along the margin on each side. They do not see well, and depend on touch, using the long setae on their pincers. When running, they hold their pincers outstretched. Males have broader pincers and longer tails than do females.

Most scorpions live in warm, dry climates. But one species is found as far north as Alberta; in Europe, *Euscorpius germanus* is found to altitudes of 1,800 m (6,000 ft.) in the southern Alps; others are found near snow in the southern Andes. Species are restricted to small areas, and only one is found world-wide in tropics. Scorpions feed at night on insects and spiders that are caught with the pincers and sometimes stung. Scorpions that remain under stones or bark during the day carry their tails to one side; burrowers hold their tails up.

In courtship the male holds the female by the pincers or jaws and leads her back and forth. Eventually he deposits a stalked package of spermatozoa (a spermatophore) and pulls the female over it. She picks up part of the capsule with an organ on her abdomen. The young

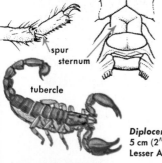

spur
sternum
tubercle

DIPLOCENTRIDAE contains tropical scorpions from the West Indies and southwestern N.A. The large *Nebo* is found in eastern Mediterranean countries. The sides of the sternum are parallel, and only one spur is present on the outside between the last leg segments. All have a tubercle underneath the tip of the stinger.

Diplocentrus hasethi
5 cm (2")
Lesser Antilles

ride on the mother's back until they shed their skins for the first time, then become independent and live a solitary life that may last several years.

Scorpions sting in self-defense. Most stings are not serious, but dangerous scorpions occur in North Africa, S.A., and Mexico. In the U.S., *Centruroides sculpturatus* of Arizona has poison that affects the nerves, causing severe pain. Species of *Centruroides* in Mexico have caused deaths. Scorpion antivenins are available.

Of over 1,200 species known, 20 to 30 occur in the U.S. Scorpions are best collected at night when they are active, with a black light, which makes them fluorescent in the darkness.

SCORPIONIDAE contains Old World and Australian scorpions. Some are giants. Only one, *Opisthacanthus*, occurs in the West Indies and Central America. The last two leg segments have only one spur on the outside, as in Diplocentridae, but there is no tubercle under the stinger.

Pandinus sp.
to 17 cm (7")
Africa; rocky areas

ventral view of sternum

Bothriurus bonariensis
5.5 cm (2.2")
southern S.A.

CHACTIDAE, represented here by *Euscorpius*, is a family world-wide in distribution. The sternum is almost square or wider than long. There may be either two spurs between the last two leg segments or one—and if one, it is on the inside. They may have two eyes on each side of the head or none. *Superstitiona*, a three-striped scorpion about 4 cm (1.6") long, is found in southwestern U.S.

BOTHRIURIDAE contains mostly South American scorpions, a few from Australia. The sternum consists of two transverse plates, much wider than long, sometimes barely visible.

VAEJOVIDAE scorpions have a broad sternum, like Chactids, but have three to five eyes on each side of the head. Some species occur in the Old World, and many are American.

sternum

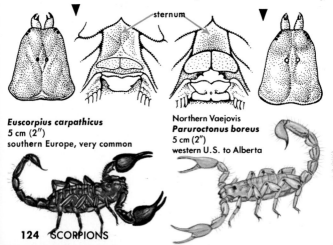

Euscorpius carpathicus
5 cm (2")
southern Europe, very common

Northern Vaejovis
Paruroctonus boreus
5 cm (2")
western U.S. to Alberta

VAEJOVID SCORPIONS

SWOLLEN-STINGER ANUROCTONUS
Anuroctonus phaeodactylus
6 cm (2.4")
Utah to California

▶

MORDANT UROCTONUS
Uròctonus mordax
6 cm (2.4")
Oregon, California

▶

STRIPED-TAILED VAEJOVIS
V. spinigerus
5 to 8 cm (2-3")
southwestern U.S.

▶

YELLOW VAEJOVIS
V. flavus
4 cm (1.5")
southwestern U.S.

▶

GIANT HAIRY HADRURUS
Hadrurus arizonensis
to 11 cm (4")
southwestern U.S.

▶

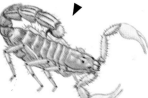

IURIDAE a small family with large,
dark tooth on base of lower jaw

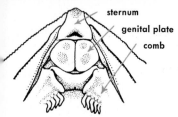

sternum
genital plate
comb

underside of *Buthus*

BUTHIDAE, with 600 species, has some representatives on all continents except Antarctica; it is the largest family of scorpions. The sternum is triangular, pointed in front and often longer than it is wide. Many species have a tubercle that is located under the stinger. The pincers have a graceful, slender hand. Some species are poisonous to man—*Tityus* in Brazil, *Androctonus* in North Africa, and *Centruroides* in Mexico.

EUROPEAN BUTHUS
B. occitanus
5 cm (2")
southern Europe;
sting painful

♀

♀

SPOTTED ISOMETRUS
Isometrus maculatus
5-7 cm (2-2.7")
cosmotropical

♂

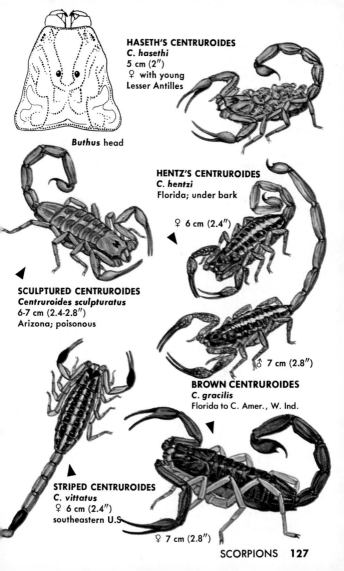

Buthus head

HASETH'S CENTRUROIDES
C. hasethi
5 cm (2")
♀ with young
Lesser Antilles

HENTZ'S CENTRUROIDES
C. hentzi
Florida; under bark

♀ 6 cm (2.4")

SCULPTURED CENTRUROIDES
Centruroides sculpturatus
6-7 cm (2.4-2.8")
Arizona; poisonous

♂ 7 cm (2.8")

BROWN CENTRUROIDES
C. gracilis
Florida to C. Amer., W. Ind.

STRIPED CENTRUROIDES
C. vittatus
♀ 6 cm (2.4")
southeastern U.S.

♀ 7 cm (2.8")

SCORPIONS **127**

HARVESTMEN (order Opiliones) are commonly called Daddy-long-legs in N.A., but many have short legs. They hold their short, round bodies near the ground. If caught by a leg, the leg may break off. The head, thorax, and abdomen are joined broadly, and in the middle of the head are two eyes, usually one on each side of a high turret. In many species, spines, tubercles, or bristles arm the head; scent glands open along the edge of the carapace. Males are smaller than females and have longer legs. Unlike other arachnids, Harvestmen do not court before mating. With a long, stout ovipositor, the female deposits her eggs in the ground in the fall; the young hatch in the following spring. Most species live less than a year or two. Harvestmen feed mainly on living insects, sometimes on dead animals or plant juices. There are 225 species in N.A., 4,500 to 5,000 world-wide.

MITE HARVESTMEN, CYPHOPHTHALMI are all less than 3 mm (0.1") long. Eyes, if present, are far apart and indistinct, and the scent glands are on cones. Unlike most mites, they have a segmented abdomen. Mite Harvestmen inhabit humid leaf litter. Probably not rare, but seldom collected. There are five families. Five species are recorded for the U.S. Captives live four to six years.

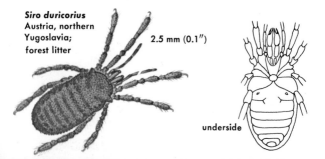

Siro duricorius
Austria, northern
Yugoslavia;
forest litter

2.5 mm (0.1")

underside

LANIATORES include about 2,000 species, mostly tropical, but with a few representatives in southern and western U.S. and in Europe. Laniatores have the basal segments (coxae) of the first three pairs of legs touching at the midline. The small family Oncopodidae (not illustrated), found only in southeastern Asia, has the body (except for the last segment) covered by a hard plate. The Assamiidae (not illustrated) carry their slender pedipalpi crossed over. Included are 300 species from southeastern Asia, Australia, and Africa.

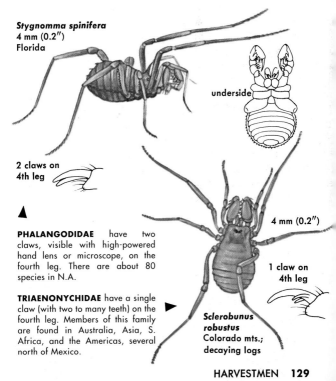

Stygnomma spinifera
4 mm (0.2″)
Florida

underside

2 claws on
4th leg

PHALANGODIDAE have two claws, visible with high-powered hand lens or microscope, on the fourth leg. There are about 80 species in N.A.

TRIAENONYCHIDAE have a single claw (with two to many teeth) on the fourth leg. Members of this family are found in Australia, Asia, S. Africa, and the Americas, several north of Mexico.

4 mm (0.2″)

1 claw on
4th leg

Sclerobunus robustus
Colorado mts.;
decaying logs

COSMETIDAE are all American species and most are tropical. All have a small median third claw at the tip of the fourth leg. The small palpi have segments flattened and keeled.

Vonones ornata
4 mm (0.2")
southeastern U.S.

underside   claws

GONYLEPTIDAE is a family of about 600 species in the American tropics. It includes the harvestmen with the largest body. All have a small median third claw at the tip of the fourth leg. The large palpi are not flattened.

Sadocus polyacanthus
11 mm (0.5")
Chile

Acrographinotus sp.
12 mm (0.5")
Peru

PALPATORES contains most temperate zone Eurasian and N.A. harvestmen. Of a total of about 900 species, 120 occur north of Mexico. Basal leg segments (coxae) separated by the breastplate (sternum).

Trogulus tricarinatus
10 mm (0.4")
Europe; under stones

TROGULIDAE are slow and nocturnal, with the eye tubercle projecting in front and overhanging the jaws. The European *Trogulus*, commonly encrusted with soil particles, is a predator of snails. *T. tricarinatus* has been established in upstate N.Y.

underside

NEMASTOMATIDAE have very short jaws and no claws on their pedipalpi. The first and fourth basal leg segments (coxae) have marginal rows of spines. The eyes are on a tubercle. Seven species occur north of Mexico, 50 in the world.

underside

Crosbycus dasycnemus
1.5 mm (0.06")
eastern Canada, U.S.

CERATOLASMATIDAE have microscopic sculpturing on legs.

Nemastoma bimaculatum
4 mm (0.2")
Europe, eastern Canada

HARVESTMEN **131**

Sabacon cavicolens
4 mm (0.2″)
eastern N.A.

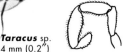

Taracus sp.
4 mm (0.2″)
western U.S.; forests   palp

SABACONIDAE usually have huge jaws, lack claws on the pedipalps, and have short legs. Nine species north of Mexico. Related Ischyropsalididae in Europe, Asia.

PHALANGIIDAE are widespread. Each pedipalp has a small claw at its tip, and in most species, the legs are long and stilt-like. Of about 800 species, 80 to 100 occur north of Mexico.

CADDIDAE have large eyes and are related.

Ischyropsalis sp.
5 mm. (0.2″)
Europe
Ischyropsalididae

Mitopus morio
♀ 5 mm (0.2″)
N. Hemisphere

Caddo sp.
1.5 mm (0.1″)
northeastern N.A.
Caddidae

Odiellus sp.
6 mm (0.2″)
N.A.

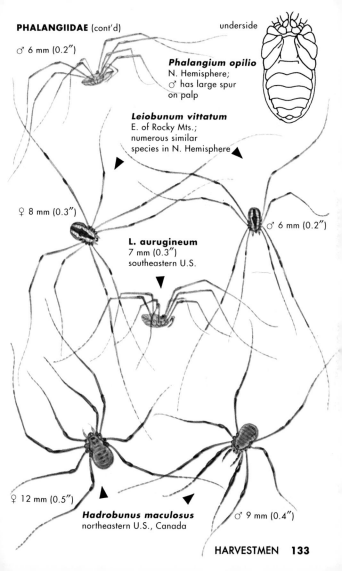

PHALANGIIDAE (cont'd)

♂ 6 mm (0.2")

underside

Phalangium opilio
N. Hemisphere;
♂ has large spur
on palp

Leiobunum vittatum
E. of Rocky Mts.;
numerous similar
species in N. Hemisphere

♀ 8 mm (0.3")

♂ 6 mm (0.2")

L. aurugineum
7 mm (0.3")
southeastern U.S.

♀ 12 mm (0.5")

Hadrobunus maculosus
northeastern U.S., Canada

♂ 9 mm (0.4")

MITES (order Acari) are found throughout the world and in all types of habitats. They surpass all other spiderlike animals in numbers. Many are parasites on plants or animals, and some transmit disease. Others are freeliving predators. Some are aquatic—mostly in fresh water, a few in the ocean. Among the parasites are some extraordinarily specialized species, adapted to live only in bird feathers or nostrils, under bat wings, in monkey lungs, bee tracheae, and similarly restricted habitats.

Most mites are very small, but they can be collected in abundance from soil or litter by using a Tullgren funnel (p. 18). A mite's body is fused into one piece, with no separation between head and abdomen. In parasites, the mouthparts may be specialized as sharp, piercing stylets. Adult mites have four pairs of legs, but larval mites have only three. Some wormlike, microscopic gall mites have fewer legs. More than 30,000 species are named, probably a fraction of the total. Many have become pests since the widespread use of pesticides.

MESOSTIGMATID MITES (suborder Mesostigmata) have either a simple claw on the pedipalps, or none. Many have plates on the back and underside. Some are free-living predators and can be collected with a Tullgren funnel. Others are parasites and live in the fur of mammals, on birds, or on insects.

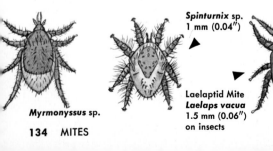

Spinturnix sp.
1 mm (0.04")

Myrmonyssus sp.

Laelaptid Mite
Laelaps vacua
1.5 mm (0.06")
on insects

TROMBIDIID MITES (suborder Trombidiformes) include many plant and animal parasites and some predators. Water Mites (p. 136) are usually placed in this group.

larva,
0.3 mm (0.01")

half grown

CHIGGER
Trombicula sp.

adult, 3 mm (0.1")

CHIGGERS, or Harvestmites (Trombiculidae), form a family of about 700 species. Adults are predators of insects or insect eggs; larval stages are parasites. Fewer than 50 species attack humans, biting where clothing is tight and causing severe itching. After feeding, they fall off. Some people are immune. In the Orient, chigger mites carry scrub typhus. In the U.S. they are most abundant in the South.

SPIDER MITES, or Red Spiders (Tetranychidae), 0.3-0.8 mm (.01-.03"), are serious pests on various crops. They often invade buildings in fall. All have silk glands opening near mouth and make a loose web among leaves.

VELVET MITES (Trombidiidae) are of little economic importance. There are several thousand species. Larvae are parasites on insects; large adults (to 4 mm, 0.2"), usually velvety red, eat insect eggs.

web with spider mites

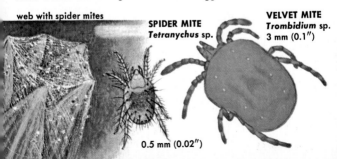

SPIDER MITE
Tetranychus sp.

0.5 mm (0.02")

VELVET MITE
Trombidium sp.
3 mm (0.1")

WATER MITES (Hydrachnellae), a group of several families, are related to trombidiid mites (p. 135). Many are colorful—red, green or yellow. Some crawl on stones, others swim. The long hairs on their legs help them to paddle, but their bodies are streamlined and nearly smooth.

Most Water Mites are predators, but some are parasitic on clams or insects. Those that parasitize dragonflies attach themselves to nymphs ready to leave the water, then to become airborne on the newly molted adult. They drop off into the water, possibly when the insect lays its eggs.

Arrenurus superior
2 mm (0.1″)

Limnochares americana
3 mm (0.1″)

Lebertia sp.
3 mm (0.1″)

SARCOPTID MITES (suborder Astigmata) includes the Scabies or Mange Mites, the Cheese Mites found in many stored products and causing the allergy "grocer's itch," and the Oribatid Mites.

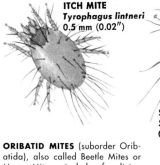

ITCH MITE
Tyrophagus lintneri
0.5 mm (0.02")

SCABIES MITE
Sarcoptes scabei
0.5 mm (0.02")

ORIBATID MITES (suborder Oribatida), also called Beetle Mites or Moss Mites, include free-living mites. With their hard, shiny shells they may resemble tiny dark brown or black beetles. Some can tuck their legs under hinged "wings," forming a little ball. Most are less than 1 mm long. In some, the shed skins collect on the back, making a high ornament. Most oribatids live in soil or litter, but those found on decaying wood are seen most often. They are of economic importance in breakdown of litter and formation of humus. Some feed on tapeworm eggs and transmit them to sheep.

Oribatula sp.
1 mm (0.04")

Phthiracarus sp.
1 mm (0.04")

walking

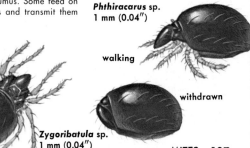

withdrawn

Zygoribatula sp.
1 mm (0.04")

well fed
♀ **WOOD TICK,**
Dermacentor, sp.
2 cm (0.8″)

TICKS (suborder Ixodida) are the largest of the mites, as much as 3 cm (1.2″) long after feeding. All are external parasites of reptiles, birds, or mammals. Most drop off their host after feeding. They molt and then wait on the tips of leaves, forelegs outstretched, ready to attach to any animal brushing past. Young ticks have only three pairs of legs. The bite of some ticks may cause mild paralysis to man; others transmit disease. Some tiny *Ixodes* sp. transmit Lyme disease. Ticks attach themselves to the host only with their mouthparts, and feed on blood. In removing a tick, take care not to leave mouthparts behind.

About 300 species of ticks occur around the world. Soft ticks (Argasidae) have a leathery integument; they have no head plate, and the head is on the underside. Hard ticks (Ixodidae) have a hard plate above the head and the head is directed forward.

SOFT TICKS

4 mm (0.2″)

5 mm (0.2″)

BLUE BUG
Argas persicus
parasite of birds
and bats; often pests
of poultry in warm, dry
parts of world

***Ornithodoros* sp.**
parasites of mammals,
including man; may
transmit relapsing
fever in western U.S.

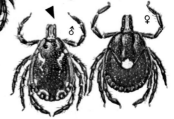

LONE STAR TICK
Amblyomma americanum
4 mm (0.2")
southcentral U.S.;
on large mammals, man

♂   ♀

WOOD TICK   4 mm (0.2")
Dermacentor andersoni
western N.A., Rocky Mts.;
attacks large mammals, man;
may transmit
diseases. *D. variabilis*
in eastern N.A.

♂ ♀

BROWN DOG TICK
Rhipicephalus sanguineus
4 mm (0.2")
found in houses,
rarely bites man;
cosmopolitan

♀

RABBIT TICK
Haemaphysalis leporispalustris
3 mm (0.1")
N.A.

♂

CASTOR BEAN TICK
Ixodes ricinus
4 mm (0.2")
cattle and sheep
pest in Eurasia

♂   ♀

CATTLE TICK ◀
Boophilus sp.
4 mm (0.2")
stays on host
throughout life

♀   ♂

HARD TICKS

MITES   **139**

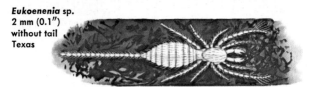

Eukoenenia sp.
2 mm (0.1")
without tail
Texas

MICROWHIPSCORPIONS (order Palpigradi) are agile arachnids less than 2 mm (0.1") long. They live under stones, going deeper into the soil if it gets too dry. Of the 50 to 60 species known, 7 occur in the U.S., in Florida, Texas, Oregon, and California; other species are found in warmer regions of the world.

RICINULEIDS (order Ricinulei) resemble ticks. They move about slowly in litter and leaf mold of humid, warm areas, and may feed on termites. They are 10–15 mm (0.4–0.6") long, with a heavy cuticle. In front of the head is a unique hinged hood covering the jaws. Young Ricinuleids have only

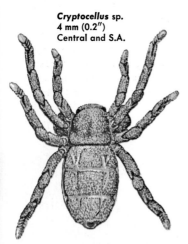

Cryptocellus sp.
4 mm (0.2")
Central and S.A.

six legs. As recently as 1929 fewer than 40 specimens of Ricinuleids were known to have been found, all from Africa and S.A. Now they are known to occur widely in the American tropics and in Mexican caves. Only one species, found in Texas, is known from the U.S. The female carries the single egg between the bent pedipalps and the hood. About 55 species are known.

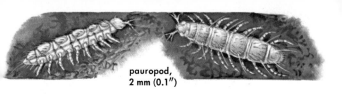

pauropod,
2 mm (0.1")

MYRIAPODS

Myriapods are the classes of arthropods in which the body is made up of numerous similar segments.

PAUROPODA, a class closely related to millipedes, are soft-bodied, less than 2 mm (0.1") long, with 12 segments and usually nine pairs of legs. The more than 500 described species are found in the tropics and temperate regions, where they live in decaying logs, forest litter, humus, and under stones. They probably feed on fungi and decaying animals. Most are sensitive to light and to drying.

SYMPHYLA, a small class of 160 known species closely related to centipedes and millipedes, are found in the tropics and temperate regions. Symphyla are 2-8 mm (0.1-0.3") long, with 15-22 back plates, 12 pairs of walking legs, and a pair of spinnerets. They live in moist, rotting wood or in soil and litter. Unlike centipedes and millipedes, they move all legs on one side together. They feed on decaying vegetation but may attack living plants and become garden pests.

GARDEN CENTIPEDE
Scutigerella immaculata
8 mm (0.3")
Europe, N.A., Hawaii;
pest in gardens
and greenhouses

CENTIPEDES (class Chilopoda) have one pair of legs on each segment. The fastest are the ones with fewest legs, the Scutigeromorphs and Lithobiomorphs. Since the plates on their back hook together, they can hold the body straight in running. The Scolopendromorphs and many-legged Geophilomorphs are slow and may move with snakelike motions; all their dorsal trunk plates are alike and move freely. About 2,500 species of centipedes are known. The orders of centipedes are described and illustrated on pp. 143–145.

Like insects and millipedes but unlike spiders, centipedes have a pair of antennae. Most are nocturnal. In daytime they hide in litter, under loose bark, stones, leaves, and debris. Some dig into the soil. They avoid extremely wet or dry niches. Centipedes are predators, feeding mostly on other arthropods. All have venom glands opening through their jaws, and the bite of even a small centipede, if it succeeds in breaking the skin, can produce pain. *Scolopendra,* the largest, can give a very painful bite, but usually is not dangerous. Some centipedes lack eyes and are blind; others do not see well. In experiments in which glass beads were dipped in "fly juice," some centipedes tried to bite the beads. They would not touch glass beads dipped in water, however, and so it is assumed that centipedes find their prey by smell and perhaps by touch.

Some centipedes can produce silk, which is used only during mating and when capturing prey. When a male finds a female, they touch antennae, and he follows her.

The male makes a small web in which he deposits a package of sperm. The sperm is then picked up by the female. The male's building of a web has been observed only among the Geophilomorphs, Lithobiomorphs, and Scolopendromorphs.

compound eye
antenna
maxilla 2
maxilla 1
poison fang
(maxilliped)
face

COMMON SCUTIGERA
Scutigera coleoptrata
3 cm (1.2")
southern Europe and southern U.S.;
outdoors. Farther north
in buildings

SCUTIGEROMORPHA is an order of mostly tropical centipedes. About 130 species have been described; all belong to the family Scutigeridae. Unlike other centipedes, they have a round head and large compound eyes. They have 15 pairs of long legs. A female of *Scutigera* produced 35 eggs over a period of days. The eggs are not guarded. Young centipedes are born with only seven pairs of legs, but the number increases with every molt. In captivity, the common *Scutigera* found throughout the U.S. may live more than a year. It is the only centipede common in houses, where it runs rapidly along walls and floors as it hunts for flies and other insects. It is surprisingly agile. In summer it lives outside, or in warm climates it lives outside all year.

fang

underside of head

Lithobius forficatus N.A., Europe

3 cm (1.2")

LITHOBIOMORPHA, or Stone Centipedes, are all less than 4.5 cm (1.7") long. Adults have 18 body segments, 15 pairs of legs and 20 to 50 or more segments in the antennae. As in Scutigeromorphs, the young have fewer legs. When disturbed, they move the last pairs of legs rapidly, throwing droplets of sticky material at the potential predator and slowing it in the tangling mass. About 1,100 species have been described in this order.

GEOPHILOMORPHA, or Soil Centipedes, are slender, eyeless centipedes that have 31 to 177 pairs of legs and antennae with 14 segments. The longest are 16-17 cm (6"), but most are less than 5 cm (2"). The number of pairs of legs is always odd, usually variable within the species, and each leg can be moved independently. Soil Centipedes penetrate as deep as 40 to 70 cm (16-28") into soil, feeding on insect larvae and worms. If disturbed, they coil up and eject from pores on their underside a secretion that seems to repel potential predators. A number of species can give off phosphorescent material. Females guard their eggs. About 1,000 species.

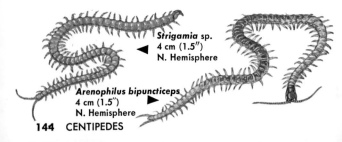

Strigamia sp.
◄ 4 cm (1.5")
N. Hemisphere

Arenophilus bipuncticeps
4 cm (1.5") ►
N. Hemisphere

SCOLOPENDROMORPHA have 21 or 23 pairs of legs and 17 to 30 segments in their antennae. World-wide and mainly tropical, the order includes the largest centipedes. Many are an attractive blue-green, olive-green, or yellow. To pick one up, it is best to use two hands and forceps as they can inflict a painful bite and can also pinch with their last pair of legs. They have been observed feeding on toads and lizards. These large centipedes dig burrows and also make chambers in which they rest. The female often coils around her eggs or young and may periodically lick the eggs, presumably to clean them. About 550 species described in two families: Scolopendridae and the blind Cryptopidae.

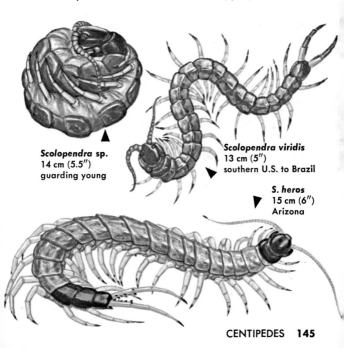

Scolopendra sp.
14 cm (5.5")
guarding young

Scolopendra viridis
13 cm (5")
southern U.S. to Brazil

S. heros
15 cm (6")
Arizona

MILLIPEDES (class Diplopoda) are "thousand leggers," differing from the "hundred leggers" or centipedes (p. 142) in having two pairs of legs on most body rings (which are two fused segments). They have one pair of antennae. Some millipedes are cylindrical, adapted for burrowing; others are flat. Some resemble woodlice (p. 152). Some are soft-bodied, very small, and covered with saw-toothed hairs (p. 147).

Millipedes are found under stones, in moist soil and leaf litter. They avoid light and feed on various plant materials, especially soft decomposing plant tissues. Most millipedes have pores, usually along the sides of their body rings, from which they discharge strong-smelling secretions that may be repellent and poisonous to other animals. Some large tropical millipedes squirt their secretions. Nevertheless, millipedes are prey of birds, toads, and other animals.

In most millipede groups, the legs on the seventh ring of the male are modified as copulatory organs (gonopods). The shape of these organs is important in identifying species. The female lays her eggs in the soil; some species construct egg capsules and guard them. Just-hatched millipedes have few rings and three pairs of legs, the number increasing with each molt. Some species live only one year, others to seven. More than 10,000 species are known, about 600 north of Mexico.

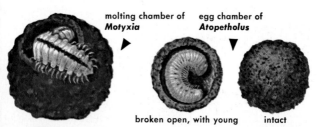

molting chamber of **Motyxia** ▶

broken open, with young

egg chamber of **Atopetholus** ▼

intact

Polyxenus lagurus
3 mm (0. 1″)
Europe, similar
species in N. A.

PSELAPHOGNATHA is a subclass of soft millipedes, all less than 4 mm (0.2″) long, with 11 to 13 rings. They have many rows and bundles of barbed hairs. Males do not have legs modified for copulation. Found in most parts of the world, there are about 60 species of two families; Polyxenidae has 5 species in N.A. They feed on algae and live under loose bark and in litter.

CHILOGNATHA, the subclass including all other millipedes, have hardened body rings and males have legs modified for copulation. There are a number of orders.

GLOMERIDA, Pill Millipedes, are short and wide, resembling woodlice (p. 152), but have more than seven pairs of legs. The male's last pair of legs are modified to hold the female when mating. Glomerids can coil into a ball when disturbed, some tropical species approaching the size of golf balls. Species found in southern states and in California are never more than 8 mm (0.3″) long. Common in Europe, where some species reach a length of 2 cm (0.8″).

Glomeris romana
11 mm (0.5″)
Italy

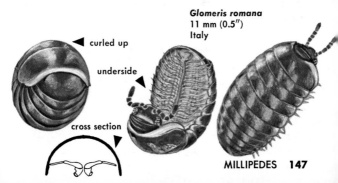

◄ curled up

underside

cross section

MILLIPEDES 147

Brachycybe sp. southern U.S.

2.5 cm (1")

1.2 cm. (0.5")

Polyzonium bikermani Arkansas

cross section

head

PLATYDESMIDA are flat with a tiny head and from 30 to 192 body rings. In males, the 9th and 10th pairs of legs are modified as copulatory organs (gonopods). The usual color is pink. The rough dorsal surface has a narrow longitudinal groove. About a dozen species occur in North America north of Mexico.

POLYZONIIDA are very similar to Platydesmida but are smooth and lack the dorsal groove. The usual color is cream. About 15 species occur north of Mexico. The orders Platydesmida and Polyzoniida, with small heads, belong in the super order Colobognatha.

CALLIPODIDA, with up to 89 rings, have a pair of tiny spinnerets on the last body ring. Most are cylindrical, with lateral ridges and knobs. In males, the second pair of legs on the 7th ring are modified as gonopods. Most produce strong-smelling white secretions from stink glands. About 20 species are known north of Mexico, one widespread in western Europe and many in eastern Mediterranean. Some are fast-running scavengers and predators.

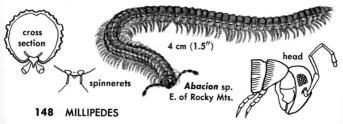

cross section

spinnerets

4 cm (1.5")

Abacion sp. E. of Rocky Mts.

head

POLYDESMIDA is an order in which most species have 20 rings that usually bear prominent keels which give them a flat-looking back. All lack eyes; the first pair of legs on the 7th ring are gonopods. Many are brightly colored; almost all have stink glands. Of 2,700 species, about 250 occur north of Mexico, fewer in Europe.

GREENHOUSE MILLIPEDE
Oxidus gracilis
2.5 cm (1″)
cosmopolitan in greenhouses and subtropics

cross section

head

Pseudopolydesmus serratus
3 cm (1.2″)
eastern N.A.

Pachydesmus crassicutis
7 cm (2.8″)
southeastern U.S.

Motyxia sp.
4 cm (1.5″)
southern California;
bioluminescent

Sigmoria aberrans
4 cm (1.5″)
N. Carolina, Virginia

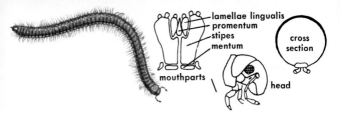

lamellae lingualis
promentum
stipes
mentum
mouthparts

cross section

head

JULIDA is one of four orders of cylindrical millipedes that have both pairs of legs on the 7th ring of the male modified as copulatory organs (gonopods). More than 100 species are found north of Mexico. The numerous members of the family Paraiulidae, or wireworms, measure 15-90 mm (.7-3.5″) long. Most are smooth, with up to 74 body rings and the gonopods are outside the body. In males, the first pair of legs are greatly enlarged. In the family Julidae, the gonopods are in a pouch, and the male's first pair of legs are hook-shaped. They are native to Europe and western Asia; introduced to parks and gardens in N.A.

SPIROSTREPTIDA is an order of large (to 28 cm, or 11″) cylindrical millipedes found mainly in the tropics. There is only one pair of gonopods, the anterior. In southwestern U.S. and adjacent Mexico, millipedes of the genus *Orthoporus* often congregate in large numbers. Three species occur in states adjacent to Mexico.

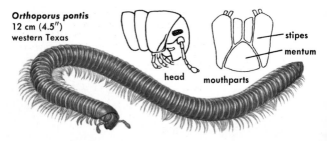

Orthoporus pontis
12 cm (4.5″)
western Texas

head   mouthparts

stipes
mentum

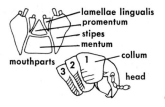

lamellae lingualis
promentum
stipes
mentum
mouthparts
collum
1
3 2
head

Cambala speobia
to 2.5 cm (1")
Texas; caves

CAMBALIDA have no legs on the 4th ring; other millipedes have one pair. *Cambala* (family Cambalidae) are easily recognized by the large cover (collum) of the first body segment and by the prominent longitudinal ridges on the body of most species. Cambalida are rare in the Great Plains and semi-deserts. The largest, to 60 mm (2.5"), occur in the Appalachian Mountains.

SPIROBOLIDA have only one pair of legs on the 5th ring; other millipedes have two pairs. The male's copulatory organs (gonopods) are hidden in a pouch. The four cylindrical groups can also be separated by the structure of the mouthparts on the underside of the head. About 35 species occur north of Mexico. *Narceus* places each single egg, 1 mm long, in a capsule of chewed leaf litter. The capsule is passed posteriorly by the legs and into the rectum where it is molded and then deposited in a pile with many others.

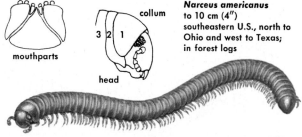

mouthparts

collum
3 2 1
head

Narceus americanus
to 10 cm (4")
southeastern U.S., north to
Ohio and west to Texas;
in forest logs

Crustacea are arthropods with two pairs of antennae. Though most are aquatic (Crayfish, Lobsters, Barnacles, Shrimps) live successfully on land. Terrestrial Amphipods (Beach Fleas, Scuds) are found on ocean beaches and in humid tropics. Terrestrial Copepods and Ostracods are found in temperate and tropical regions. Terrestrial Isopods (Woodlice) are found in a variety of habitats.

WOODLICE (order Isopoda, suborder Oniscidea) usually feed on dead plant material. Over 100 species north of Mexico. Females carry eggs and young in a brood pouch.

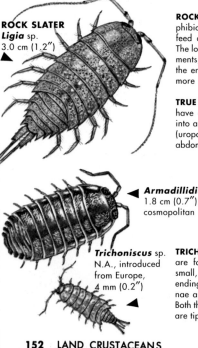

ROCK SLATER
Ligia sp.
3.0 cm (1.2")

ROCK SLATERS (Ligiidae) are amphibious on ocean beaches. They feed at night, mainly on seaweed. The long antennae have 5 basal segments, followed by a distinct bend; the ending segments number ten or more in this family.

TRUE PILL BUGS (Armadillidiidae) have an arched body and can roll into a ball when disturbed. The tails (uropods) are shorter than the last abdominal segment.

► *Armadillidium vulgare*
1.8 cm (0.7")
cosmopolitan

► *Trichoniscus* sp.
N.A., introduced
from Europe,
4 mm (0.2")

TRICHONISCIDS (Trichoniscidae) are found in wet spots. They are small, narrow-bodied, and have the ending segments of the long antennae appearing as one section only. Both the antennae and tails (uropods) are tipped by a brush.

WOODLICE from most other families cannot roll into a ball. Oniscids have three ending segments following the bend in the long antennae. Porcellionids and trachelipodids have two ending segments following the bend in the long antennae.

Oniscus asell
1.6 cm (0.6")
N.A., introduced
from Europe

Porcellio scaber
1.7 cm (0.7")
widespread, N.A.,
introduced from
Europe tubercles
all over

**Porcellionides
pruinosus**
1.2 cm (0.5")
cosmopolitan
body dusty-looking

Trachelipus rathkei
1.5 cm (0.6")
cosmopolitan;
in ♂, 3rd segment
from end of 7th leg
has keel

CYLISTICIDS (Cylisticidae) have an arched body but unlike true pill bugs, the tails (uropods) stick out when the animal rolls up.

Cylisticus convexus
1.5 cm (0.6")
N.A., introduced
from Europe;
shiny

LAND CRABS (Gecarcinidae), found only in the subtropics and tropics, are land-dwelling crustaceans, but the females return to the ocean to reproduce. They dig tunnels 30-40 cm deep (12-16"), 18 cm (7") in diameter, and come out at night to feed. Little ones climb walls and trees.

Land Hermit Crabs (Coenobitidae) can give a good pinch with their colorful claw if handled carelessly. Most are scavengers. In the Southwest Pacific, the Coconut Crab *(Birgus),* growing to 45 cm (18") in length, feeds on fallen coconuts, and can be destructive to crops. It is considered a delicacy itself.

LAND HERMIT CRAB
Coenobita clypeatus
claw to 6 cm (2.3") diameter
eastern Caribbean, southern Florida

LAND CRAB
Gecarcinus lateralis
to 9 cm (3.5") wide
Florida Keys, Bermuda,
West Indies

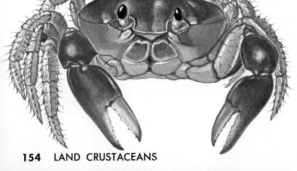

MORE INFORMATION

BOOKS

Dondale, C. D., and J. H. Redner, *Spiders* in **The Insects and Arachnids of Canada,** parts 5, 9, 17, 19, Research Branch, Agriculture, 1978–1992.

Foelix, R. F., **Biology of Spiders,** Oxford University Press, New York, 1996. Spider biology with emphasis on physiology and behavior.

Kaston, B. J., **Spiders of Connecticut,** rev. ed., Bul. Conn. Geol. Nat. Hist. Surv. 70, 1981. The most useful reference book on spiders of the eastern U.S.

Polis, G. A., ed., **The Biology of Scorpions,** Stanford University Press, Stanford, CA, 1990.

Roberts, M. J., **Spiders of Britain and Northern Europe** (Collins Field Guide), Harper Collins, London, UK, 1995. A beautifully illustrated volume on European spiders.

Roth, V. D., **Spider Genera of North America,** 3rd. ed., Amer. Arachnology Soc., Dept. of Zoology, University of Florida, Gainesville, FL, 1994. Key to families and genera found in North America.

Ruppert, E. E., and R. D. Barnes, **Invertebrate Zoology,** 6 ed., Saunders, Philadelphia, 1994. A good textbook giving background on Crustaceans, Myriapods, and Arachnids.

Shear, W. A., ed., **Spiders: Webs, Behavior, and Evolution,** Stanford University Press, Stanford, CA, 1986.

Weygoldt, P., **Biology of Whip Spiders (Chelicerata: Amblypygi),** Apollo Press, Copenhagen, Denmark, 2000.

Wise, D. H., **Spiders in Ecological Webs,** Cambridge University Press, Cambridge, UK, 1993.

Woolley, T. A. **Acarology, Mites and Human Welfare,** Wiley, New York, 1988.

WEB SITES

The Internet is full of interesting information about spiders. Some keywords to use in searches include arachnology and spiders. You can also find information by searching for a particular spider, such as Black Widow, Jumping Spider, or Tarantula.

157

MEASURING SCALE (IN MILLIMETERS AND CENTIMETERS)